# To Connect is to Understand Mathematics 4
## Selected works 1981-1998

ALFINIO FLORES

Copyright © 2017 Alfinio Flores

All rights reserved.

ISBN: 1975803175
ISBN-13: 978-1975803179

# DEDICATION

To all my mathematics teachers who made plenty of connections and gave me the opportunity to make additional connections.

# CONTENTS

| | | |
|---|---|---|
| | Acknowledgments | vii |
| 1 | The kinematic method and the Geometer's Sketchpad in geometrical problems | 1 |
| 2 | Mean machines | 11 |
| 3 | Ancestry of humans and bees | 19 |
| 4 | Curves as envelopes with the Geometer's Sketchpad | 27 |
| 5 | Fibonacci in the forest | 38 |
| 6 | Sí se puede. It can be done: Quality mathematics in more than one language | 46 |
| 7 | Tin-can ice cream | 58 |
| 8 | Bilingual lessons in early-grades geometry | 65 |
| 9 | Connections in proportional reasoning: Levers, arithmetic means, mixtures, batting averages, and speeds | 79 |
| 10 | Orchestrating, promoting, and enhancing mathematical discourse in the middle school: A case study | 94 |
| 11 | Geometry of numeric iterations | 111 |
| 12 | Connections: A lottery, a computer and the number $e$ | 120 |
| 13 | The shadows of mathematics | 127 |
| 14 | Pythagoras meets Van Hiele | 130 |
| 15 | Mathematics for gifted students in grades K-3: Early intervention and identification of gifted potential | 143 |
| 16 | A geometrical approach to mathematical induction: Proofs that explain | 153 |
| 17 | Mathematical connections with a Spirograph | 164 |

| | | |
|---|---|---|
| 18 | Three approaches to the golden ratio | 175 |
| 19 | Calculators in calculus: that's the limit | 179 |
| 20 | A puzzle of mathematical formulas | 185 |
| 21 | Estimation Performance and Strategy Use of Mexican 5th and 8th Grade Sample | 189 |
| 22 | Calculus for middle school teachers using computers and graphing calculators | 212 |
| 23 | They're off! | 218 |
| 24 | Exploration of the mean as a balance point | 225 |
| 25 | Family Planning | 232 |
| 26 | Parabella | 236 |
| 27 | A Microcomputer and the law of small numbers | 243 |
| 28 | Otagato | 246 |

# ACKNOWLEDGMENTS

The articles were published before in the journals, proceedings, and books indicated. The complete reference to the original source is given in a footnote in the corresponding page. I appreciate the free and prompt permissions received to republish these articles from these publishers:
*International Journal of Computers for Mathematical Learning* (Springer)
*School Science and Mathematics* (Wiley)
*Mathematics and Computer Education*
*PRIMUS* (Taylor and Francis)
*SCOPE*
*Educational Studies in Mathematics* (Springer)
*Providing a foundation for teaching middle school mathematics* (State University of New York Press)
*Proceedings Third Annual Conference on Technology in Collegiate Mathematics*

Thanks to the direct intervention of NCTM's president Matt Larson I was granted permission by NCTM in 2016 to include in this collection my articles from the journals *Mathematics Teacher, Teaching Children Mathematics, Arithmetic Teacher* and the books *Multicultural and gender equity in the mathematics classroom, Calculators in Mathematics Education, Projects to Enrich School Mathematics Level 1.*

I want to thank my coauthors of articles in this volume: Isabel Perkins, Lori Birge, Audra Guest, Judy Sowder, Randy Philipp, Bonnie Schappelle, Rosa Isela Pérez, Doug McLeod, Barbara Reys, Robert Reys, and Arthur White. It was a pleasure to work with them. Their names appear in the reference to the published version.

The version of the articles included in this collection stems from my own files. In some cases, they do not include the last editorial improvements from the publishers.

# 1 THE KINEMATIC METHOD AND THE GEOMETER'S SKETCHPAD IN GEOMETRICAL PROBLEMS[1]

**Introduction**

To solve geometric problems, sometimes it is useful to imagine how the elements of a figure will change as some of its elements move. The Geometer's Sketchpad provides a very powerful aid to help students see how some of the elements change as the others are "dragged." How some elements depend on others can be made evident and the solution of the problem will be clear. In some cases, the relations between magnitudes of the segments, the angles, etc. in a geometric figure are not as easy to see as the relations between the velocities of variation of these magnitudes as the figure is deformed.

The capacity of the Sketchpad to trace the locus of points by drawing the position of the points at different times as they move provides a powerful help to see how the velocities are related. Of course, the best way to get a feeling is by actually moving points around with the Sketchpad, but one example may help the reader see what is meant.

As one point is dragged, if we select the Trace locus option, the Sketchpad will draw successive positions of the point. If the point is dragged quickly, the separation between the points drawn will be bigger than when the point is dragged slowly. In figure 1, the magnitude of the velocity of C is half that of B as both are dragged in a loop simultaneously. The separation between successive positions of B is twice as big as that of C. The total distance traveled by B will also be twice that of C. In this case the two trajectories are similar, with a dilation factor of 2.

---

[1] Flores, A. (1998). The kinematic method and the Geometer's Sketchpad in geometrical problems. *International Journal of Computers for Mathematical Learning, 3,* 1-12. Used by permission of Springer.

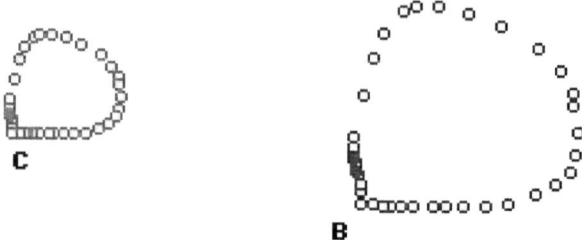

Figure 1. Velocities of points

In this paper we will consider the points as endpoints of changing vectors. We will think of segments of geometrical figures that are deformed as changing vectors. As some geometrical figures are deformed, some of the relations between its parts will change, while others will remain invariant. In some cases, by proving invariance of the relations between velocities of endpoints we will be able to prove the invariance between the relation of parts of the figure.

Of course, by using the Sketchpad, in many cases the student will see the relations between the desired parts and can measure to verify them. This is an important aspect of conjecturing. However, measuring does not always provide a clue as to how to prove the results. In this paper, we will provide some examples of how by looking at the velocities of variation we can find the relations between the magnitudes by using the theory of velocities, that is, the kinematic method. The problems presented can, of course, be solved using many different strategies and tools. It is important, however, to encourage our students to find alternative ways to solve problems.

**Relations between the vectors and the velocities of their endpoints.**
The reader should be familiar with the basic results of vector algebra. Let $\mathbf{r} = \mathbf{r}(t)$ be a vector function, with pole O and endpoint M. If at a time $t_0$ the vector is equal to $\mathbf{r}_0$, and at $t_1$ it is equal to $\mathbf{r}_1$, then the vector $\Delta\mathbf{r} = \mathbf{r}_1 - \mathbf{r}_0$ will be the change of $\mathbf{r}$ in the time interval $\Delta t = t_1 - t_0$. The average velocity of the endpoint M for that time interval is given by $\Delta\mathbf{r} / \Delta t$. The instantaneous velocity of the endpoint M of the vector $\mathbf{r}$ at time $t_0$ is given by $\mathbf{v} = \lim_{\Delta t \to 0} \frac{\Delta \mathbf{r}}{\Delta t}$.

**Two theorems**
The following results that we present without proof will be used in this paper (for proofs see for example Lyúbich & Shor, 1978). Let $\mathbf{r}, \mathbf{r}_1, \mathbf{r}_2$ be vectors with pole O and endpoints M, $M_1$ and $M_2$ respectively. The respective velocities of the endpoints will be $\mathbf{v}, \mathbf{v}_1$ and $\mathbf{v}_2$.

<u>Theorem 1a</u>. Let $\mathbf{r}_1, \mathbf{r}_2, \mathbf{r}$ be vectors. If the endpoints of the vectors move in

such a way that for all time $\mathbf{r} = \mathbf{r}_1 + \mathbf{r}_2$, then their velocities are related by the corresponding equation $\mathbf{v} = \mathbf{v}_1 + \mathbf{v}_2$.

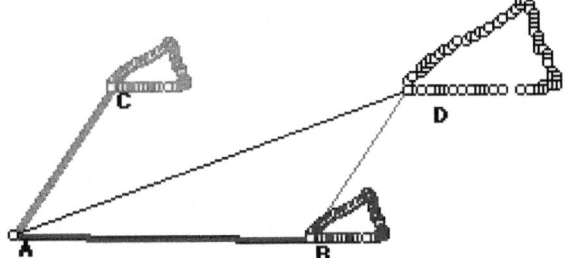

Figure 2. Sum of vectors and relation of velocities.

<u>Theorem 1b</u>. If the velocities $\mathbf{v}_1$, $\mathbf{v}_2$, $\mathbf{v}$ of the endpoints three vectors $\mathbf{r}_1$, $\mathbf{r}_2$, $\mathbf{r}$ satisfy for all time the relation $\mathbf{v} = \mathbf{v}_1 + \mathbf{v}_2$, then the vectors satisfy for all time the relation
$\mathbf{r} = \mathbf{r}_1 + \mathbf{r}_2 + \mathbf{k}$, where $\mathbf{k}$ is a constant vector.

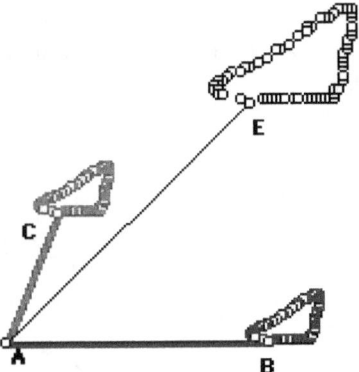

Figure 3a. Sum of velocities.

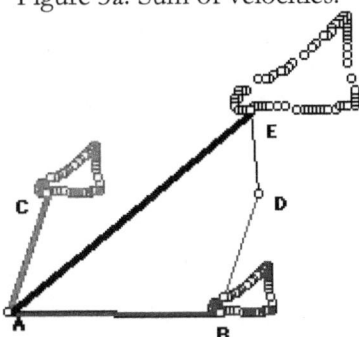

Figure 3b. Sum of velocities and relation of vectors.

For theorems 2a and 2b, let $U_\alpha \mathbf{a}$ be the rotation of vector $\mathbf{a}$ by an angle $\alpha$.

Theorem 2a. If the endpoints of vectors $\mathbf{r}_1$, $\mathbf{r}_2$ change in such a way that for all time $\mathbf{r}_1 = mU_\alpha \mathbf{r}_2$, where m is a constant number and $\alpha$ is a constant angle, then their velocities of change are related for all time by the equation $\mathbf{v}_1 = mU_\alpha \mathbf{v}_2$.

Figure 4. $AD = 1/2\ U_{\pi/2}\ AB$

Theorem 2b. If the velocities $\mathbf{v}_1$ and $\mathbf{v}_2$ of the endpoints of the vectors $\mathbf{r}_1$ and $\mathbf{r}_2$ are related for all time by the equation $\mathbf{v}_1 = mU_\alpha \mathbf{v}_2$ where m is a constant number and $\alpha$ is a constant angle, then the vectors are related for all time by the equation $\mathbf{r}_1 = mU_\alpha \mathbf{r}_2 + \mathbf{k}$ where $\mathbf{k}$ is a constant vector.

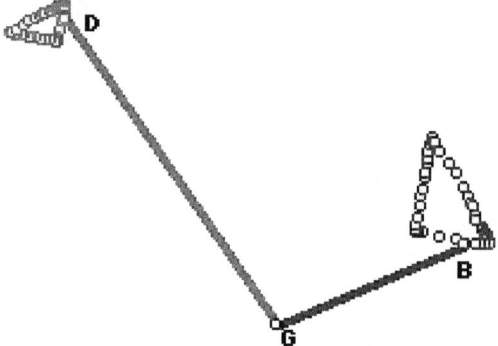

Figure 5. $\mathbf{v}_1 = 1/2\ U_{\pi/2}\ \mathbf{v}_2$.

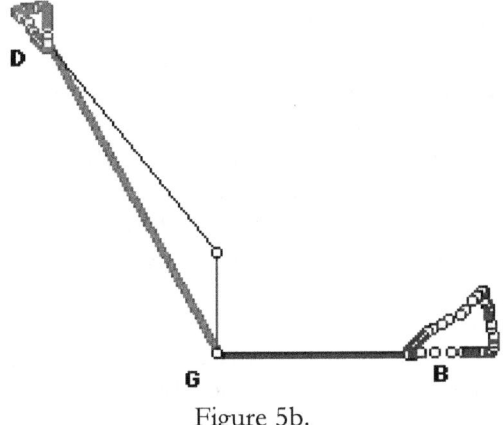

Figure 5b.

## Example 1. Treasure hunt

Gamow relates the story of a man who finds the instructions to find a treasure on an island. From the gallows walk to the oak counting the steps. At the oak turn right by 90° and take the same number of steps. Put a spike on the ground. Return to the gallows and walk towards the pine counting the steps. At the pine turn left by 90° and take the same number of steps. Put a second spike here. The treasure is in halfway between the spikes (see figure 6). The man finds the island, the oak and the pine, but to his dismay the gallows are gone, no trace left. In despair, he starts digging everywhere, but to no avail, the island is too big.

Students can simulate the situation using the Geometer's Sketchpad, putting the gallows at an arbitrary place, and observe what happens to the location of the treasure as the position of the gallows is changed.

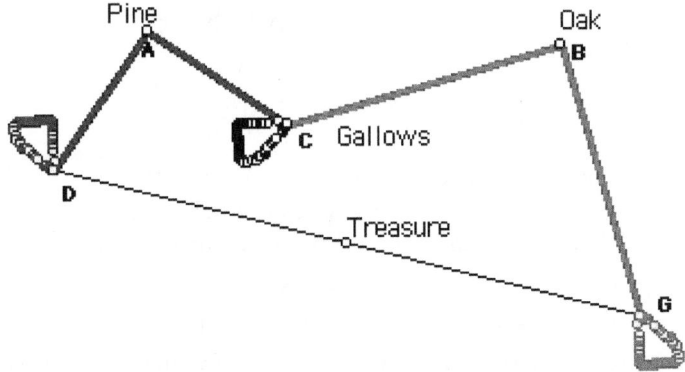

Figure 6. Position of the treasure

We can also use the kinematic method to show that the position of the treasure is fixed. Let C move with velocity $v_0$, and let $v_1$ and $v_2$ be the velocities of D and G. For all time $AD = U_{-\pi/2}AC$. Therefore, by theorem 2a, $\mathbf{v}_1 = U_{-\pi/2}\mathbf{v}_0$. Similarly, because $BG = U_{\pi/2}BC$, $\mathbf{v}_2 = U_{\pi/2}\mathbf{v}_0$. Therefore, the velocity of G is the same as the velocity of D rotated by 180°, that is, $\mathbf{v}_1 = -\mathbf{v}_2$. The velocity of its midpoint is $\mathbf{v} = (\mathbf{v}_1 + \mathbf{v}_2) / 2 = 0$, which shows that the point corresponding to the treasure does not move. In other words, the position of the gallows does not matter. Starting from any point on the island, following the instructions we can find the treasure. (Don't rush to find where the island is; Gamow omitted the location of the island, and changed the names of the trees to keep the secret).

**2)** Let AGE and ABC be two isosceles triangles with right angles at A, and that share that vertex (see figure 7). Then BG and EC are congruent and perpendicular. To see this, let B move with velocity $\mathbf{v}_1$ and C with velocity $\mathbf{v}_2$. For all time $AC = U_{\pi/2}AB$. Therefore, by theorem 2a $\mathbf{v}_1 = U_{\pi/2}\mathbf{v}_2$ for all time. By theorem 2b $EC = U_{\pi/2}GB + \mathbf{k}$ for all time. To see that $\mathbf{k} = 0$, we can choose a convenient position, for example, let B coincide with A.

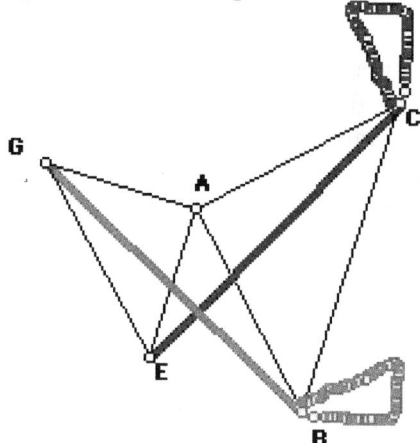

Figure 7. $GB = U_{\pi/2}EC$

**3)** Construct squares on sides AB and AC of a triangle ABC. Let D be the midpoint of BC. Let H and M be the centers of the squares (see figure 8). Then HD and MD are congruent and perpendicular. To see this let A, H, and M move with velocities $\mathbf{v}$, $\mathbf{v}_1$, and $\mathbf{v}_2$. Notice that $BA = \sqrt{2}U_{-\pi/4}BH$ for all time. Therefore, $\mathbf{v} = \sqrt{2}U_{-\pi/4}\mathbf{v}_1$. Notice also that $CA = \sqrt{2}U_{\pi/4}CM$ for all time. Therefore, $\mathbf{v} = \sqrt{2}U_{-\pi/4}\mathbf{v}_2$. Thus, $\mathbf{v}_1 = U_{\pi/2}\mathbf{v}_2$ for all time. By theorem 2b, $DH = U_{p/2}DM + \mathbf{k}$, where $\mathbf{k}$ is constant. To see that $\mathbf{k} = 0$, we can choose a convenient position, for example, let A coincide with D.

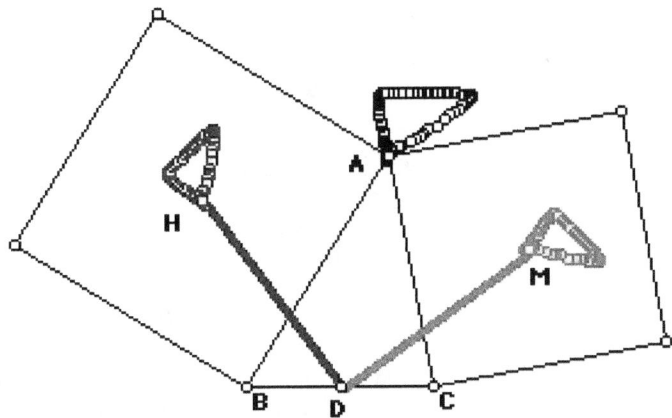

Figure 8. DH = $U_{\pi/2}$ DM

**4)** Construct exterior squares on the sides of a triangle ABC. Let G, Q, L be the centers of the squares as shown in figure 9. Then GQ is congruent and perpendicular to AL.

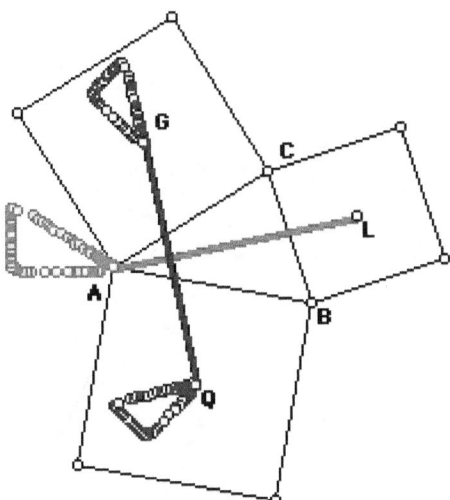

Figure 9. LA = $U_{\pi/2}$ GQ

To see this, let A move with velocity **v**, G and Q with velocities $\mathbf{v}_1$ and $\mathbf{v}_2$. CA = $\sqrt{2} U_{\pi/4}$ CG for all time, and BA = $\sqrt{2} U_{-\pi/4}$ BQ for all time. Therefore, **v** = $\sqrt{2} U_{\pi/4} \mathbf{v}_1$, and **v** = $\sqrt{2} U_{-\pi/4} \mathbf{v}_2$, that is, if **v**, $\mathbf{v}_1$ and $\mathbf{v}_2$ have a common pole, the pole and the endpoints of the vectors form a square, and **v** is the diagonal (see figure 10). The magnitude of the velocity of G with respect to Q is given by the other diagonal of the square, which therefore has the same length and is perpendicular to the velocity **v**. Therefore, GQ = $U_{\pi/2}$AL + **k**, where **k** is constant. To see that **k** = 0 choose a convenient position, for example when A coincides with C.

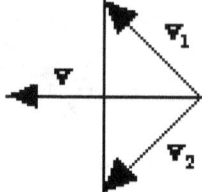

Figure 10. Diagonal vectors on a square.

**5)** On a quadrilateral construct exterior squares on its sides (see figure 11). The lines connecting the centers of squares of opposite sides are congruent and perpendicular. To see this, move segment AB parallel to itself with velocity **v**. Comparing DA with DK, and CB with CT, we can see in the same way as the previous example that the velocity of T relative to K is equal in magnitude and perpendicular to the velocity **v**. Therefore, PX = $U_{\pi/2}$TK + **k**. To see that **k** = 0, make A coincide with D, which will be the situation of example 4).

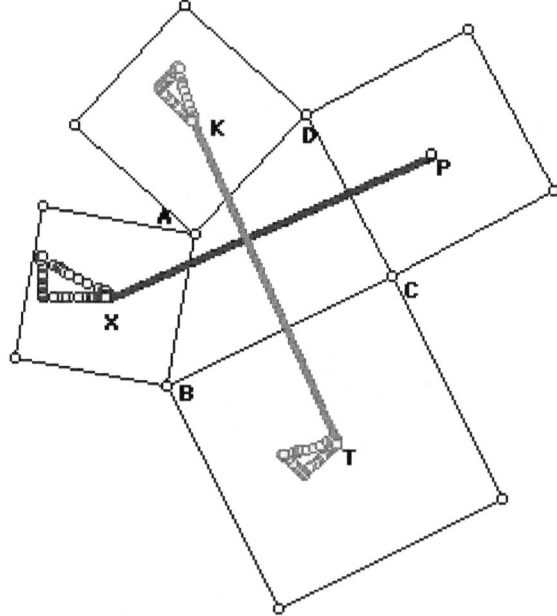

Figure 11. PX = $U_{\pi/2}$KT

### Exercises

1) Torricelli's triangles. Construct exterior equilateral triangles on the sides of a triangle ABQ (see figure 12). Show that the velocities of D and G have the

same magnitude and that they differ by an angle of 60°. (Hint: compare QA with QD, and BA with BG). Use theorem 2b to conclude that VD = $U_{\pi/3}$VG + **k**. Show that **k** = 0 by choosing a convenient position for point A.

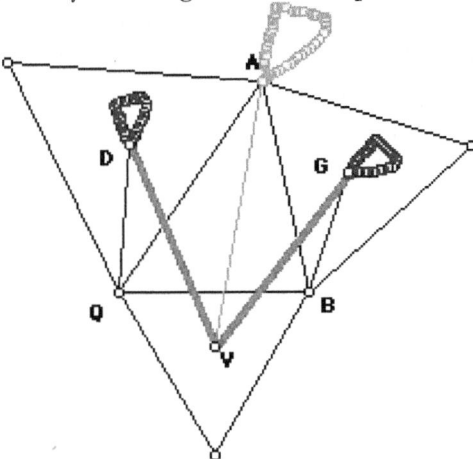

Figure 12. VD = $U_{\pi/3}$VG

2) On a parallelogram construct exterior squares on its sides (see figure 13). Show that the centers of those squares form a square. Hint: Drag segment AC parallel to itself and trace the locus of the centers Q, G, N. The velocity of Q will be the same as the velocity of the segment. Show that the velocity of G is the same as the velocity of N rotated by a 90° angle. Therefore, JN = $U_{\pi/2}$JG + **k**. Show that **k** = 0 by choosing a special position (for example when the parallelogram is a square).

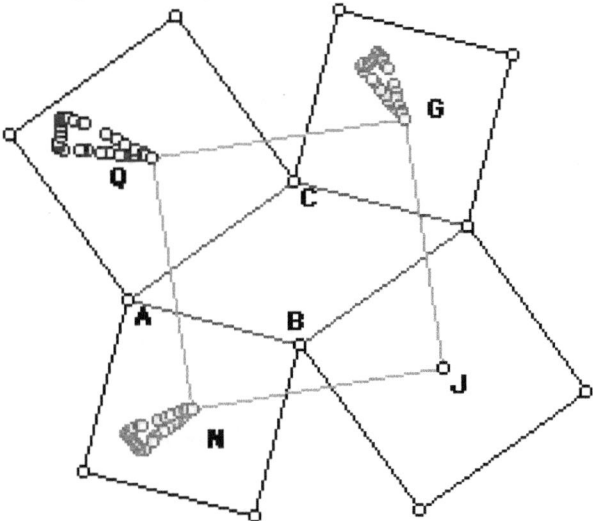

Figure 13. JN = $U_{\pi/2}$JG

**References**

Gamow, George. (1972). *One two three ... infinity*. New York: Bantam.

Lyúbich, Yu. I., and Shor, L. A. (1978). *Método cinemático en problemas geométricos*. [Kinematic method in geometrical problems]. Moscow, Mir.

# 2 MEAN MACHINES[2]

One of the goals emphasized by the *Standards* (NCTM, 1989) is that students establish connections between different topics of mathematics. Having students explore alternative representations of numerical concepts and operations is one way to help them establish such connections. Nomographs and slide rules are tools that have been used historically to compute (Glenn & Johnson, 1973). In this article we invite students to think about why these devices work and explore other mathematical concepts with them.

**The arithmetic mean machine**
Scales that have numbers forming an arithmetic sequence equally spaced on three parallel lines can be used to find the arithmetic mean (or average) of two numbers. Simply locate the numbers to be averaged, one on each of the outer scales, join them with a straight line, and find the point where this line intersects the central scale. Figure 1 shows that 3 is the average of 4 and 2. Having tried this with several pairs, students can explore more ideas. They can discuss what happens when both numbers are even, or both are odd, or when one number is even and the other is odd. They can describe what is the average when one of the numbers is 0. Students can also use the machine to obtain the average of positive and negative numbers. An activity that students enjoy is finding all the pairs of numbers that have 0 as their average. In a ninth-grade classroom, students described the resulting patterns using expressions like "it looks like a spider web," "it seems like beams of light," "there are many triangles."

*Why does it work?* When students highlight the line segments that correspond

---

[2] Flores, A. (1998). Mean machines. *Mathematics Teacher, 91*, 266-268. Copyright National Council of Teachers of Mathematics. Used by permission.

to the two numbers being averaged, the line connecting them, and the line connecting the zeros (see figure 2), they find a trapezoid. Students see that in a trapezoid the length of the mid parallel is the average of the lengths of the parallel sides, which explains why this nomograph averages.

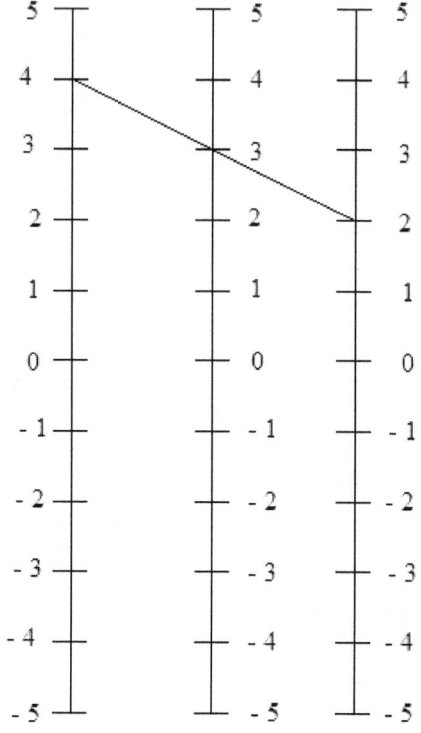

Figure 1. An arithmetic mean machine.

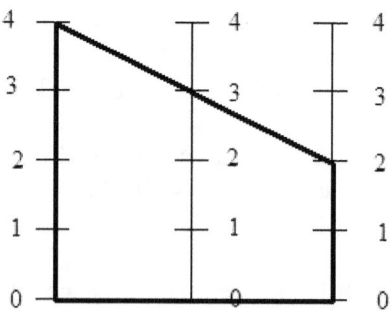

Figure 2. A trapezoid is the key for the arithmetic mean machine.

## An adding machine

By doubling the central scale, the three equidistant parallel lines can be converted into an adding machine as shown in figure 3 for the sum 4 + 2 = 6. Students may again describe the lines, their scales, and distances. They can contrast this machine with the arithmetic mean machine: How are they alike? How are they different? Students who have contrasted the central scale with the previous machine have said, "spaces between numbers are smaller," "they are half as big", "numbers grow twice as fast." As well as using the adding machine to obtain sums of pairs of numbers, students can also explore questions such as: How is the commutative property reflected in the addition machine? What numbers have 0 as their sum? What happens when a number is added to itself? How can the adding machine be used to subtract numbers? When one number is positive and the other negative, how do their distances from zero affect whether the result is positive or negative? Students can also explain why the adding machine works. The key idea is that the numbers on the central scale grow twice as fast, so the intersection of the line connecting the number with the central line is twice the average.

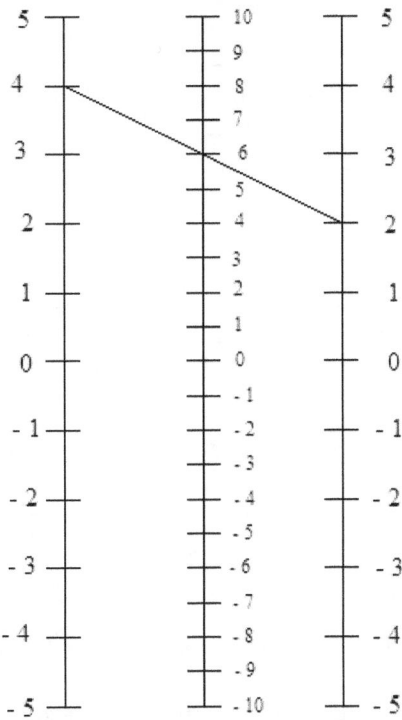

Figure 3. An adding machine.

## A multiplication machine

If we use terms of a geometric sequence on the lines, we can transform our machine into a multiplier. The spacing in the central scale is different from the spacing in the outer scales. Figure 4 illustrates the product $16 \times 4 = 64$. Students should describe the multiplication machine in their own words, describe the lines, the spacing in them, the numbers used to label the points. They should notice that all the numbers are powers of two, and that each number is twice the previous number. As well as using the multiplication machine to obtain products of numbers, students can also explore questions like: What numbers have 1 as their product? How is a number squared? How can the multiplication machine be used to divide numbers? How can we use this machine to extract square roots? Students can compare the patterns and findings with those obtained with the adding machine.

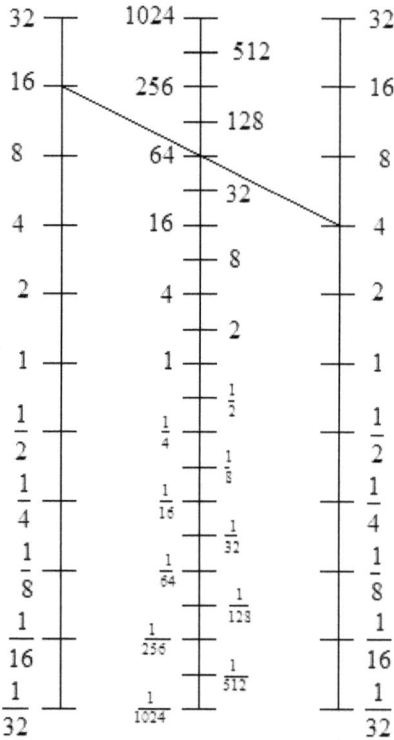

Figure 4. A multiplication machine.

*Why does it work?* To understand why the multiplication machine works, we can write the numbers on the scales as powers of 2.

# TO CONNECT IS TO UNDERSTAND MATHEMATICS 4

| ... | $\frac{1}{32}$ | $\frac{1}{16}$ | $\frac{1}{8}$ | $\frac{1}{4}$ | $\frac{1}{2}$ | 1 | 2 | 4 | 8 | 16 | 32 ... |
|---|---|---|---|---|---|---|---|---|---|---|---|
| ... | $2^{-5}$ | $2^{-4}$ | $2^{-3}$ | $2^{-2}$ | $2^{-1}$ | $2^0$ | $2^1$ | $2^2$ | $2^3$ | $2^4$ | $2^5$ ... |

To multiply two numbers of a geometric sequence, we just is add their exponents and then find the corresponding term. This multiplication machine is simply an addition machine for the exponents.

**Another mean machine (geometric mean)**
In many situations in mathematics another kind of mean is useful. For example, if we have a rectangle of sides $a$ and $b$, the length of the side of the square with the same area is $\sqrt{a \times b}$. This number is called the geometric mean of $a$ and $b$. Again, by using terms of a geometric sequence, this time equally spaced on the three lines, we can get a machine to obtain the geometric mean of two numbers (see figure 5). For example, if we join 16 and 4 on the outer scales with a straight line, the intersection of the transversal with the middle line is at 8. Students can use the geometric mean machine to obtain the geometric mean of different pairs of numbers, and they can explore patterns as they did with the other machines and contrast them.

*Why does it work?* The geometric mean of two numbers in a geometric sequence corresponds to the number whose exponent is the arithmetic average of the exponents. For example, given 2 and 32, the geometric mean is $\sqrt{2 \times 8} = 8$. Because $2 = 2^1$, and $32 = 2^5$, we can write this as $(2^1 \times 2^5)^{1/2} = (2^{(1+5)})^{1/2} = 2^{(1+5)/2}$. Thus, this is an arithmetic mean machine for the exponents.

The multiplication machine and the geometric mean machine are based on a logarithmic correspondence. Base 2 was used in this case. The slide rule, a popular device to compute before hand-calculators, was also based on a logarithmic correspondence, base 10.

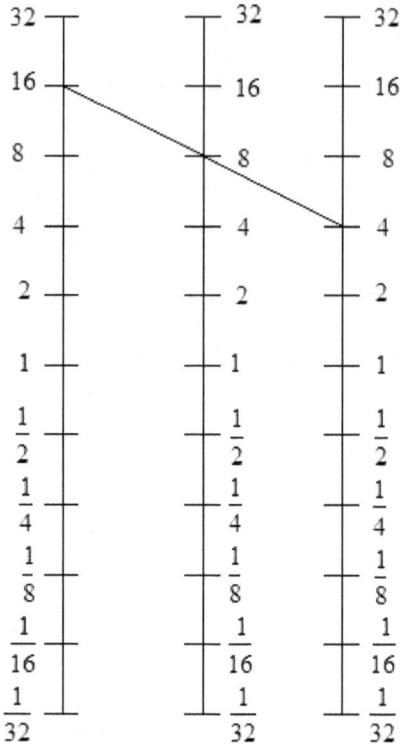

Figure 5. A geometric mean machine.

**Extensions.**
Students can make their own machines using unmarked scales (see figure 6). They can use other numbers for the intervals, for example fractions for the addition machine. They can also use other geometric sequences for the multiplication machine, for example powers of 3. They can explore how to multiply or estimate products for numbers that are not explicitly marked on the scales.

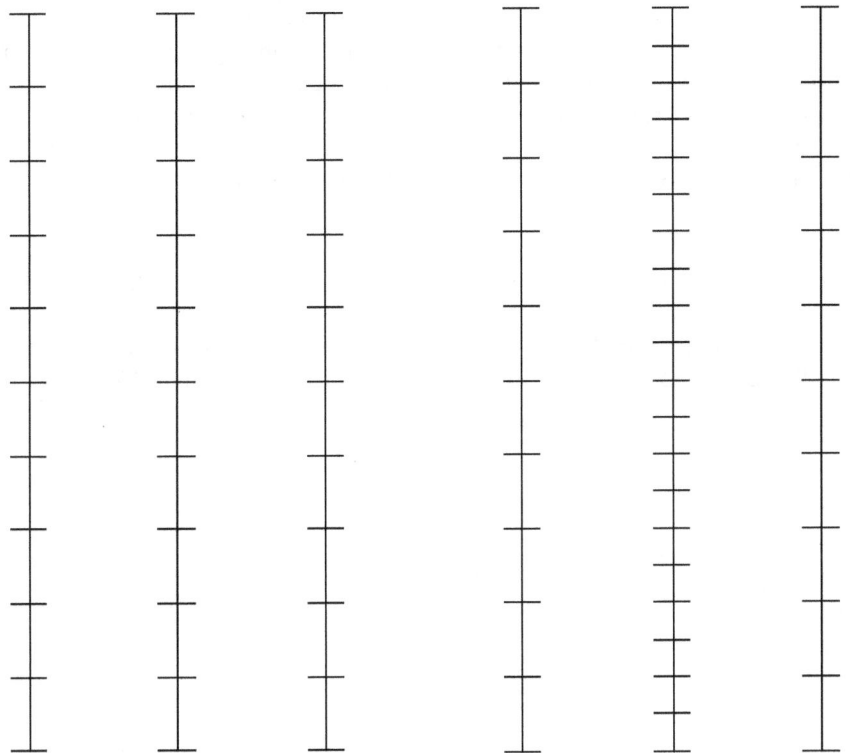

Figure 6. Unmarked scales

They can also use parallel lines to build their own slide rules to do addition (figure 7), or multiplication (figure 8) or extract cubic roots or higher. In each case, they should be able to explain their device.

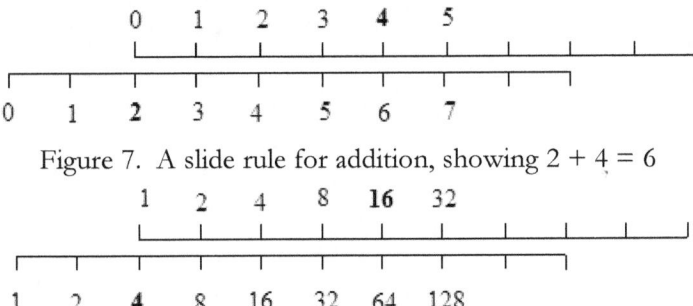

Figure 7. A slide rule for addition, showing 2 + 4 = 6

Figure 8. A slide rule for multiplication, showing 4 × 16 = 64

**Conclusion.**
The mean machines can be used to highlight the relation of the arithmetic

mean to trapezoids, giving a connection between numbers and geometry. They also provide an opportunity to realize that the group of integer numbers with the operation of addition has essentially the same structure as the group formed by the terms of a geometric sequence with the operation of multiplication. Students can thus see a concrete example of number sets with different operations that have the same algebraic structure.

**References**

Glenn, William H. & Johnson, Donovan A. *Invitation to mathematics*. NY: Dover, 1973.

National Council of Teachers of Mathematics. *Curriculum and Evaluation Standards for School Mathematics*. Reston, VA: The Council, 1989.

# 3 ANCESTRY OF HUMANS AND BEES[3]

*Grade Level:* 7-12

*Mathematics Skills/Concepts:*
Sequence, geometric sequence, ratio, powers of 2, first differences of a sequence, Fibonacci sequence, and golden ratio.
*Science Concepts/Processes:*
Sex determination, X and Y chromosomes, haploid, diploid.

*Prerequisite Skills:*
Students should know the definition of ratio, and be able to compute the ratio of consecutive terms of a sequence.
Basic knowledge about bees (differences and relations between workers, queens, and drones). Students should be familiar with the process of determination of sexuality in human beings in terms of an X chromosome received from the mother, and either an X or Y chromosome received from the father.

*Objectives:*
Compare the differing number of ancestors of humans and bees which results from a dissimilar process of sexual determination. Study and compare the corresponding sequences of numbers (Fibonacci and geometric) that describe the numbers of ancestors of humans and bees, and learn some of their properties.

---

[3] Flores, A. and Birge, L. (1998). Ancestry of humans and bees. *School Science and Mathematics, 98,* 99-103. Used by permission of Wiley and School Science and Mathematics Association.

*Rationale*

*Content background*

Biology. A person's genotype determines whether that individual is a male or female. Every organism contains genetic material in all of its cells. The genetic material found in cells is deoxyribonucleic acid (DNA) and is organized in the form of chromosomes. Sex determination in most species is determined by two chromosomes called the X and Y chromosomes. In most animals, an individual with both an X and a Y chromosome is considered to be male, while someone with two X chromosomes is considered to be female. An individual's sex is determined by the father since he can pass down either his X or his Y to his offspring, while a mother always passes down one of her two X chromosomes.

Bees, wasps, and ants have a different method of sex determination. Fertilized eggs develop into females while unfertilized eggs develop into males. The male bee begins life with only one set of chromosomes. A cell with only a single set of chromosomes is called haploid, while a cell with two sets is called diploid. Most normal body cells are diploid. Haploid or triploid (three sets of chromosomes) would not be viable in most species. The cells of male bees do not remain haploid, however. Although dividing cells appear to be haploid, chromosome duplication occurs in the absence of cell division in some, but not all tissue. This leads to the presence of two copies of each chromosome in those cells. (Normally, chromosomal duplication only occurs during cell division.)

Bee eggs receive one set of chromosomes from the queen bee. Those eggs that are fertilized receive a second set of chromosomes from a male (the father) and the larvae become female bees. Those eggs that are not fertilized have the original set of chromosomes from the queen and develop into male bees. Note that unfertilized eggs in most species would be nonviable. Later, some of the cells in the male bees that have one set of chromosomes will undergo the chromosomal duplication described above and then have a second set of chromosomes. The second set that arises through duplication will be identical to the original set.

Mathematics. A sequence of the form $a, ak, ak^2, ak^3, \ldots$ is called a geometric sequence with constant ratio $k$. In the sequence $1, 2, 4, 8, \ldots, a = 1$, and $k = 2$. The constant ratio is the quotient determined by dividing the new term by the preceding term.

The Fibonacci sequence $1, 1, 2, 3, 5, 8, \ldots$ is defined by letting the first two terms be equal to 1; the terms after the first two are obtained as the sum of the preceding two terms. The ratios of successive terms of the Fibonacci sequence, such as 8/5, 13/8 approximate closer and closer the number $\phi = 1.6180339\ldots$. This number, called the golden ratio, can be illustrated the by means of a rectangle that has the following property. If the larger square is cut out from the rectangle, the remaining rectangle is similar to the original (see Figure 1).

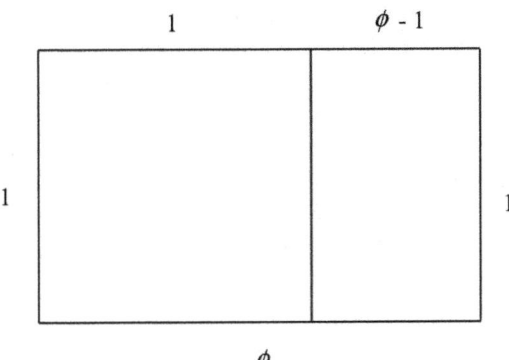

**Figure 1.** *Golden rectangle.*

That is, $\phi$ satisfies the proportion $\frac{1}{\phi} = \frac{\phi-1}{1}$. From here we can obtain the equation $\phi^2 - \phi - 1 = 0$. The positive root of this equation is $\phi = \frac{1+\sqrt{5}}{2}$.

Interrelationship of content. A situation that arises in biology, counting the number of ancestors in each preceding generation for humans and bees gives rise to two different number sequences. Those sequences are then studied and compared from a mathematical point of view.

*Research background*
Connections is one of the standards that run across all grade levels and across other content standards in the *Curriculum and Evaluation Standards for School Mathematics* (National Council of Teachers of Mathematics, 1989). According to this standard, students should use and value the connections between mathematics and other disciplines. The *National Science Education Standards* (National Research Council, 1996) emphasize unifying concepts and processes such as models and measurement. Mathematical models such as the number sequences used in this activity can help students to model and count the number of ancestors for humans and bees.

## Lesson Outline

*Time needed:* Two sessions of 45 - 50 minutes
*Materials:* Graph paper, pencil, calculator.

**Procedure**

*Part One. Ancestry of humans.*
Activity 1. Finding the number of ancestors of a human. Every human being has two parents. Each contributes one set of chromosomes to the process of forming a new human being. Human sexuality is determined by a chromosome received from the mother, which is always an X, and by one received from the father which can be either X or Y. If the combination is XX it's a girl, if it is XY, it's a boy. So, in the case of humans the number of ancestors is given by the sequence: 1 person has 2 parents, 4 grandparents, 8 grand-grandparents etc. We can represent this situation in a diagram (see Figure 2).

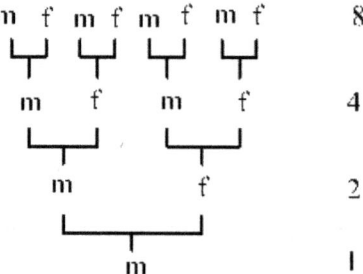

**Figure 2.** *Ancestry of a human*

In the following activity we will study the sequence 1, 2, 4, 8, 16, ..., where each term is double the preceding one. In real life, the number of ancestors of human beings does not quite follow this sequence because marriage of distant relatives is quite common.

Activity 2. Express the relation between a term and its succeeding term using a recursive relation. What is the ratio of consecutive terms? Express the terms of the sequence using powers of 2. Write the sequence 1, 2, 4, 8, ... etc. and compute the first differences between the terms. What do you observe?

```
  1     2      4      8      16      32      64      128 ...
     1      2      4      8      16      32      64 ...
```

What would be the sequence given by the second differences?
The relation between a term of this sequence and the preceding terms can also be expressed additively. Each term differs from the sum of all the previous terms by 1. For example, 8 - 1 = 1 + 2 + 4
    16 - 1 = 1 + 2 + 4 + 8

Activity 3. How would you extend the sequence to the left side? Use the fact that every term is half as big as the one to the right.

... 1/8  1/4  1/2  1  2  4  8  16...

What would be the sum of all the terms to the left of 1? Figure 3 may provide you a clue. The total area of the square is equal to the sum of the terms 1/2 + 1/4 + 1/8 + ...

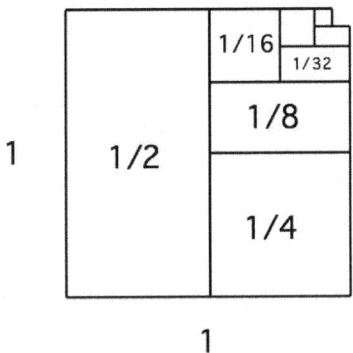

**Figure 3.** *The sum* 1 = 1/2 + 1/4 + 1/8 + ...

Using this fact, we see that with the complete sequence, every term is exactly the sum of all the terms smaller than itself.

*Part Two. Ancestry of bees.*
Activity 4. In the case of the bees, the drone (the male bee) is born from an unfertilized egg and only has one parent, his mother. He receives one set of chromosomes from her. A female bee is born from a fertilized egg and has two parents, a mother and a father. She receives one set of chromosomes from each parent. Thus, the ancestry for a drone is given by a different diagram (see Figure 4).

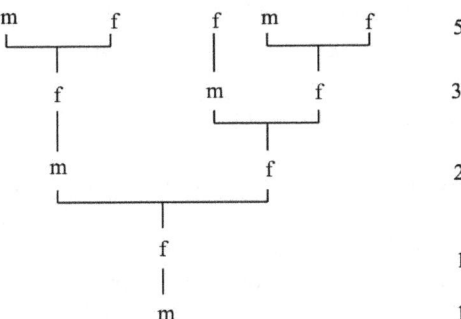

**Figure 4.** *Ancestry of a male bee.*

Figure 4 has the property that the number on each level is the sum of the two previous levels. This sequence 1, 1, 2, 3, 5, 8, 13, 21, ... was first studied by Leonardo de Pisa, in a different problem, that of a population of rabbits. Leonardo lived in 13th century Italy, and he is also known as Fibonacci. The numbers in the sequence are therefore called Fibonacci numbers.

Activity 5. Look at the first differences of the Fibonacci numbers
1   1   2   3   5   8   13   21...
  0   1   1   2   3   5   8...
Notice that you obtain essentially the same sequence. What would be the sequence of the second differences? Add all the terms of the sequence, up to a certain point. Notice that the sum differs by one from a number that is also a Fibonacci number. Which number is it?
1 + 1 + 2 + 3 + 5 + 8 = 21 - 1
1 + 1 + 2 + 3 + 5 + 8 + 13 = 34 - 1
How would you expand the Fibonacci sequence to the left?

Activity 6. Use your calculator to compute the ratios of successive Fibonacci terms (present term divided by the previous term): 1/1, 2/1, 3/2, 5/3, 8/5, 13/8, 21/13, ...
Continue until the number on your display does not change. What value does the sequence of ratios of Fibonacci numbers approach? (1.61803...). This number is called the golden ratio, and denoted by $\phi$. Use your calculator to obtain its multiplicative inverse. Observe that the decimal expansion of $1/\phi$ after the decimal point is the same 0.61803.... Therefore,

$\phi = 1 + 1/\phi$. From this equation we can obtain the proportion $1 / \phi = (\phi - 1) / 1$, which is an alternative way to define $\phi$. For relatively large terms of the Fibonacci sequence, the ratio of two successive terms is almost equal to $\phi$. Therefore, if we multiply a term of the Fibonacci sequence by $\phi$ and round the result to the nearest integer we will get the next Fibonacci number. For example, $13 \times \phi = 21.034442... \approx 21$
$$21 \times \phi = 33.978714... \approx 34$$

Activity 7. Summary comparison of the two sequences
*Ancestry of humans*
The number of ancestors in a given generation is described by $2n$.
Geometric sequence, ratio 2.
First differences give same sequence.
Sum of first $n$ terms differs by 1 from $(n+1)$th term.

*Ancestry of bees*

The number of ancestors is given by the Fibonacci sequence, which has the property that each term is the sum of the previous two terms.
First differences give same sequence.
"Almost" a geometric sequence, ratio between terms approximates $\phi$.
Sum of first $n$ terms differs by 1 from the $(n + 2)$th term.

## Evaluation

Teacher interacts with students as they develop and explore the sequences that represent the number of ancestors for humans and bees. Students can describe the mathematical properties of the sequences in their own words. Students can describe the differences from a genetic point of view between how sex is determined in humans and bees. Students can then explain how those differences are reflected in the different number sequences that arise.

## Extensions

Biology. Students can further research the effect that a closely related group of chromosomes could have in humans. For example, any heterozygous defective gene would have a good chance of being expressed. They can think about what would happen if almost everyone in the population carried a defective gene for a disease like cystic fibrosis. Your chance of getting a defective copy from both parents would be 25% if they each carried one copy. In a genetic system like bees, those odds would be increased. In male bees, the odds would go up to 50% because you would only get chromosomes from the females and you would get either the normal or defective gene. Students can discuss the implications and differences for bees in terms of number of many defective genes, immunity to mutations, or size of population and losing a number of defective offspring.

Mathematics. The geometric sequence $2^n$ grows very quickly. Students can investigate this sequence using their calculators. They can recall the legend of the invention of chess, where the prize asked was one grain of wheat for the first square, two for the second, four grains for the third, and so on, doubling the number of grains for each successive square, until all 64 squares were accounted. Students can describe this number of grains in terms of tons, yearly production of wheat in the world, etc. to get a sense of its size. Students can also study other geometric sequences where the ratio between terms is another number.

Students can also explore additional properties of the Fibonacci sequence (see for example Huntley (1970), Garland (1987), Boles and Newman (1990), and Kappraff (1991)). Related to the property that the Fibonacci sequence is "almost" a geometric sequence is the fact that each term is given by Binet's formula

$$F_n = \frac{1}{\sqrt{5}}\left[\left(\frac{1+\sqrt{5}}{2}\right)^n - \left(\frac{1-\sqrt{5}}{2}\right)^n\right]$$

Students can verify that the formula gives the values for the corresponding Fibonacci terms for $n = 1$ and $n = 2$. With a little algebraic manipulation, they can also verify that the formula satisfies the fundamental Fibonacci relationship that each term is the sum of the two previous ones. Students can also arrive at $\phi$ by looking at a sequence that is both geometric, and that satisfies that each term is the sum of the two previous terms. A geometric sequence (except for constant factors) is 1, $x$, $x^2$, $x^3$, etc. If the terms in this sequence satisfy that the sum of the first two is the third one, that is if $1 + x = x^2$, then $x$ will also satisfy the condition for other terms, $x^n + x^{n+1} = x^{n+2}$. Students can see that $\phi$ is a solution of the equation $1 + x = x^2$.

## References

Boles, M. and Newman, R. (1990). *The golden relationship: Art, math and nature.* Bradford, MA: Pythagorean Press.

Farnsworth, M. W. (1978). *Genetics.* New York: Harper & Row.

Garland, T. H. (1987). *Fascinating Fibonaccis.* Palo Alto, CA: Dale Seymour.

Huntley, H. E. (1970). *The divine proportion.* New York, NY: Dover.

Kappraff, J. (1991). *Connections: The geometric bridge between art and science.* New York, NY: McGraw Hill.

National Council of Teachers of Mathematics. (1989). *Curriculum and Evaluation Standards for School Mathematics.* Reston, VA: Author.

National Research Council. (1996). *National Science Education Standards.* Washington, DC: National Academy Press.

# 4 CURVES AS ENVELOPES WITH THE GEOMETER'S SKETCHPAD[4]

## Introduction

The envelope of a family of curves is the curve that is tangent to all the curves in the family. When the curves in the family are lines or circles, the computer can be a powerful tool for students to explore curves defined as envelopes. This article will show how software like the Geometer's Sketchpad (Jackiw, 1995) with its dynamic capabilities and its feature to trace points, lines, or circles can be easily used by students in high school and the early years of college to construct familiar curves such as the ellipse, the parabola, and the hyperbola as envelopes. Students will also be able to construct easily other wonderful curves such as the astroid, deltoid, cardioid, limaçon, and nephroid as envelopes of families of lines or circles.

## An example of envelope in real life.

A supersonic airplane flies above the ground parallel to the earth surface. For each point on the path of the airplane, the sound will expand in all directions forming a sphere. If the airplane is flying faster than the speed of sound, the airplane will be out of the sphere. All spheres formed this way are homothetic, with the airplane being the point of homothecy. All spheres will be tangent to a cone. This cone is the envelope of the spheres. The inside of the cone is the zone of audibility.

---

[4] Flores, A. (1997). Curves as envelopes with the Geometer's Sketchpad. *Mathematics and Computer Education, 31*(1), 56-65.

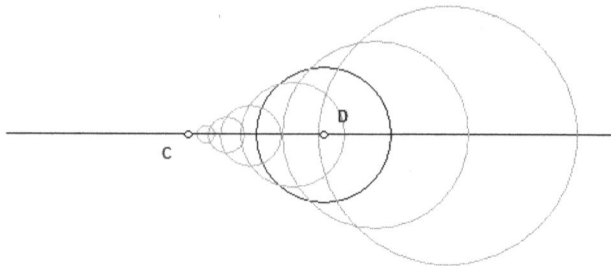

Figure 1

We can also focus what happens in a two-dimensional plane. The intersection of each sphere with a plane parallel to the line of flight is a circle. The circles obtained so on the plane have a hyperbola as their envelope. This is so because the intersection of the cone with a plane will be the envelope of the collection of all circles so formed, and the intersection of a cone with a plane parallel to the axis of the cone is a hyperbola. (For an alternative demonstration that the envelope of this family of circles is a hyperbola see for example Boltianski, 1977.)

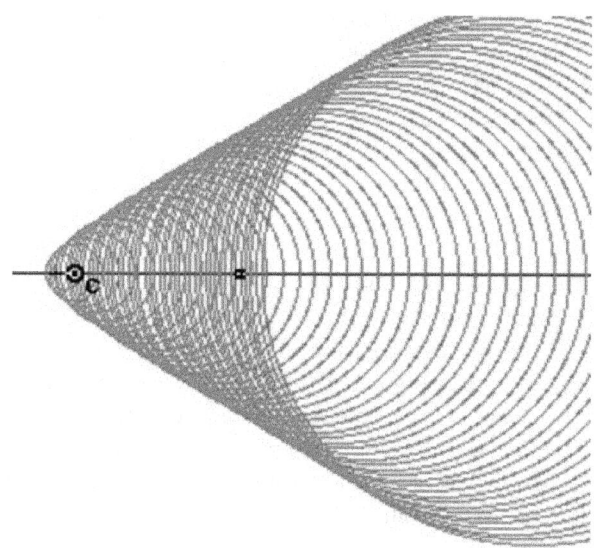

Figure 2

**Folding the conics**

The conics as envelopes of lines can also be illustrated by folding wax paper (see for example Yates, 1974, or Johnson, 1995). A fixed point C on a sheet of paper is folded over upon a point D on a fixed circle or line, and creased. The crease will be the perpendicular bisector of segment CD (see figure 3)

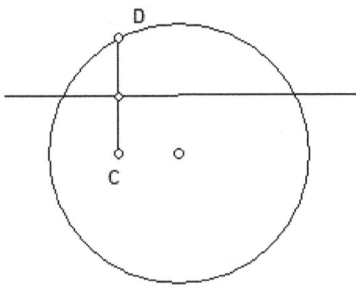

Figure 3

Now repeat the process with the same fixed point on another point of the circle or line. The creases so obtained will be part of a family of lines the point is inside the circle we obtain an ellipse, if the point is outside the circle a hyperbola will be obtained. When using a line instead of a circle we get a parabola. We can also use the Geometer's Sketchpad to simulate the creases of the paper and obtain the conics as envelopes as it will be shown below.

**Ellipse** as envelope of lines
Choose a fixed point C inside a circle. Choose a point D on the circle. Construct segment CD and find its midpoint. Construct a line through the midpoint perpendicular to the segment. Trace the line as you drag D around the circle. The family of lines so obtained will have an ellipse as envelope. What is point C? Change the position of point C. Drag D around the circle again. How does the shape of the ellipse change when the point C is closer to the center of the circle? What is the shape when C is closer to the circle?

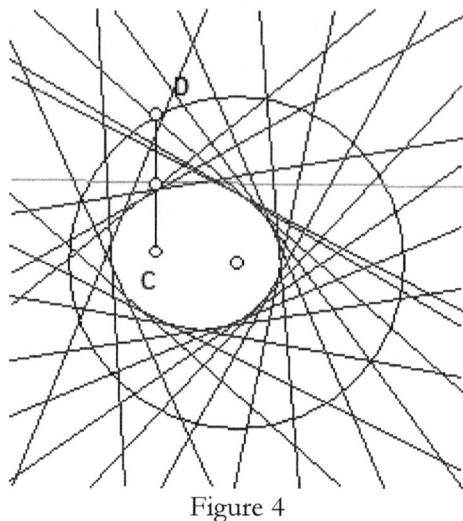

Figure 4

**Parabola** as envelope of lines
Let C be a fixed point not on a fixed line, and let D be a point on the line. Construct the segment CD and its midpoint. Construct the line perpendicular to segment CD through its midpoint. Trace this line as you drag point D on the line. The family of lines so obtained will have a parabola as envelope. What is the role of the fixed point C? What is the role of the fixed straight line? The similarity with the case of the ellipse can be highlighted if we think the line as a "circle" of infinite radius.

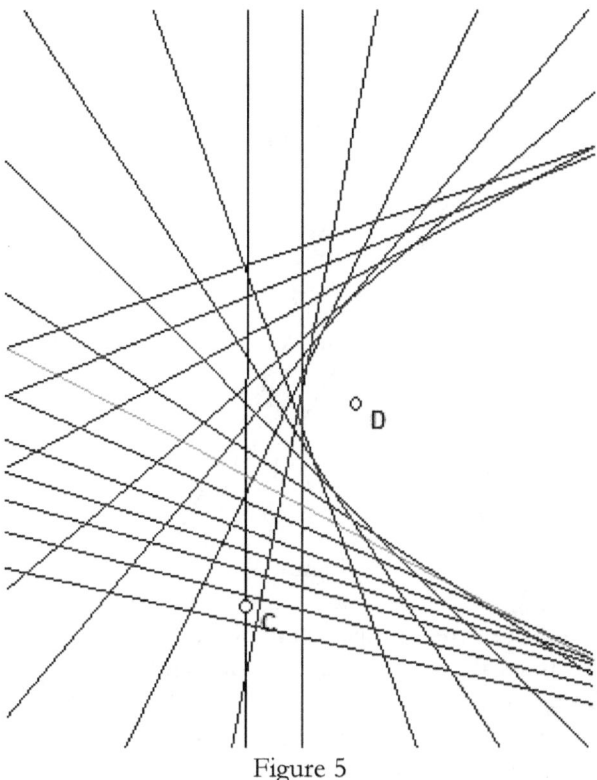

Figure 5

**Hyperbola** as envelope of lines
Choose a fixed point C outside a fixed circle. Let D be a point on the circle. Construct segment CD and its midpoint. Construct a line perpendicular to CD through its midpoint. Trace this line as you drag D around the circle. The family of lines so obtained will have a hyperbola as envelope.
Change the position of point C, and drag D around the circle. How does the shape change when C is further away from the circle?

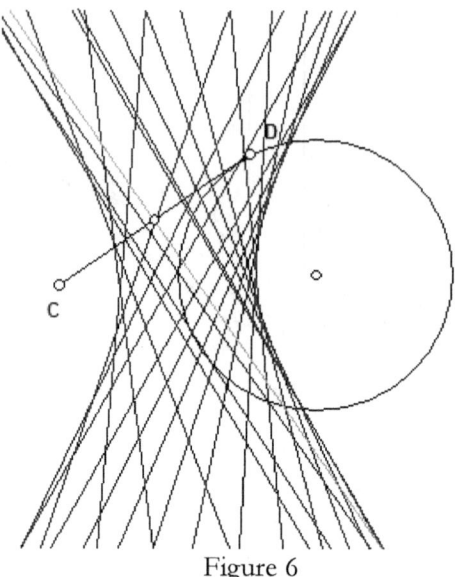
Figure 6

## Other ways to obtain the conics as envelopes

If in the instructions above we change the perpendicular line so that instead of passing through the midpoint it passes through its endpoint on the circle (or the line), we will obtain also the conics as envelopes of the corresponding families of lines (Lockwood, 1976). The following figures show examples of ellipse, parabola, and hyperbola.

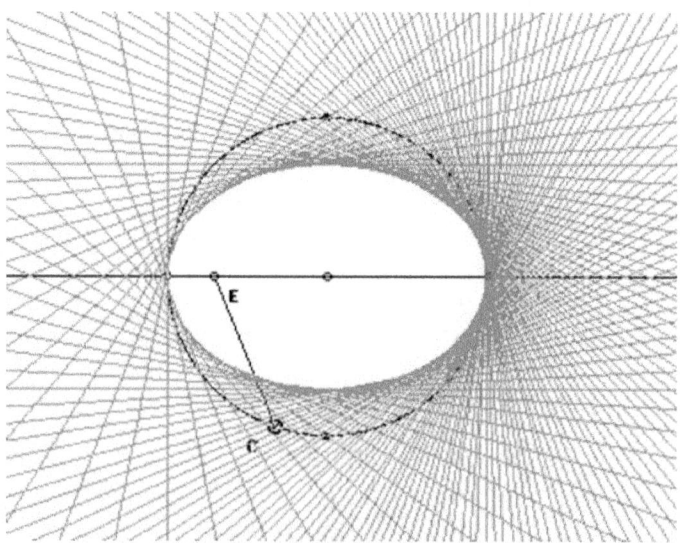
Figure 7

# TO CONNECT IS TO UNDERSTAND MATHEMATICS 4

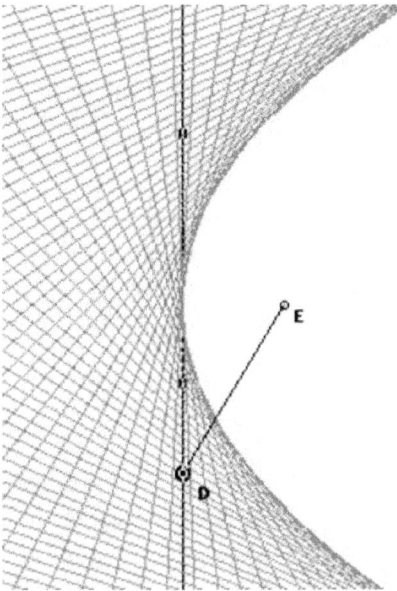

Figure 8

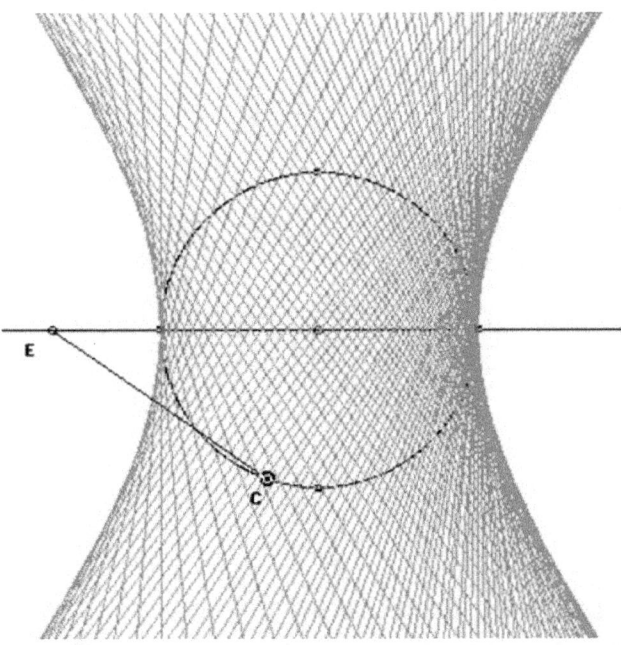

Figure 9

## Other curves as envelopes of lines

Students can also obtain other curves as envelopes using simple instructions on the computer using the Geometer's Sketchpad. This can be a first introduction to beautiful curves such as the astroid, and the deltoid. Students could later explore these curves from a different point of view, as hypocycloids using a device such as the spirograph or by using equations (see Lockwood, 1976). Some curves can be easily generated as envelopes of circles, such as the cardioid, the limaçon, and the nephroid. Students who want to learn more about these curves can consult the books by Lockwood (1976) or Yates (1974).

## Astroid as envelope of lines

E is a point on the vertical axis, H and J are points on the horizontal axis so that EH and EJ are of fixed length m. Drag E. Trace segments EH and EJ as you drag E. The segments so obtained will have an astroid as envelope.

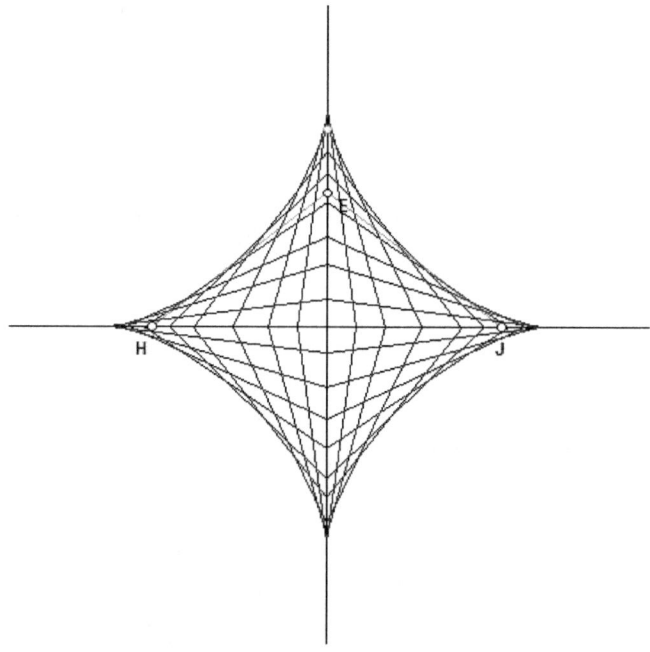

Figure 10

## Deltoid as envelope of Simson lines

Let CDE be a triangle inscribed in a circle. Let F be a point on the circle. Construct the perpendiculars from F to the three (extended) sides of the triangle. The feet of the perpendiculars are on a straight line (this in itself is a remarkable result.) This line is called the Simson line of F with respect to the triangle. Trace this line as you drag F around the circle. The envelope of the family of Simson's lines is a deltoid.

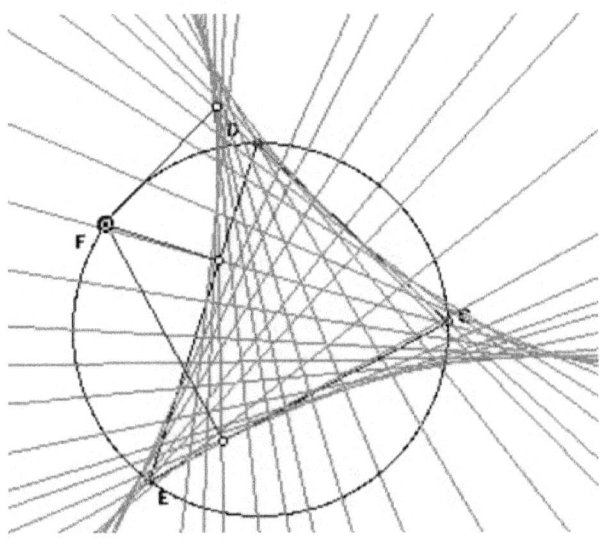

Figure 11

**Nephroid** as envelope of lines
Let I be a point on a line passing through the center A of a circle. With center I and radius IA draw a circle that intersects the original circle at K and J. The envelope of the segments IJ, and IK as I moves along the line is the nephroid.

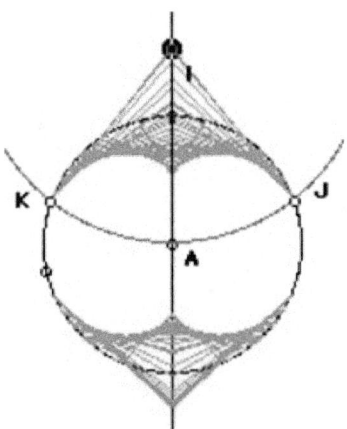

Figure 12

**Cardioid** as envelope of circles
Let C be a fixed point on a circle. Let D be another point on the circle. Construct a circle with center at D that passes through C. Trace this circle as you drag D around the fixed circle. The envelope of the circles so obtained will be a cardioid.

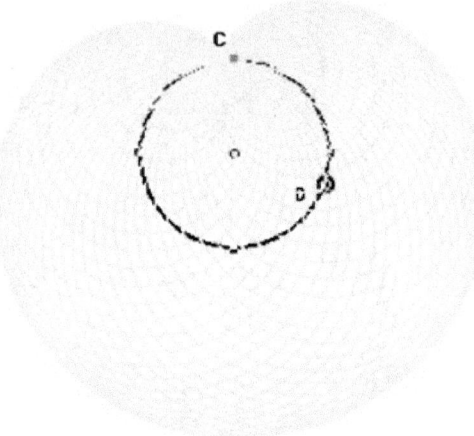

Figure 13

**Limaçon** as envelope of circles
D is a fixed point. E is a point on a fixed circle. Draw a circle with center E and radius ED. Trace this circle as you drag E around the fixed circle. The envelope of the moving circles will be a limaçon.

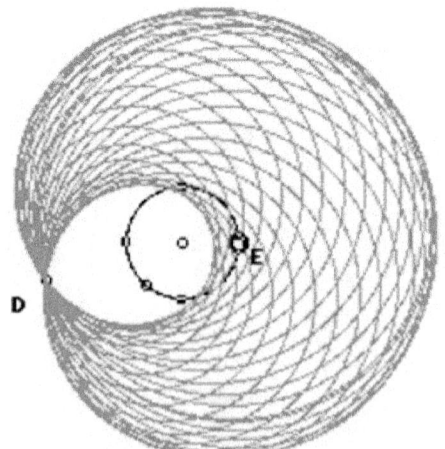

Figure 14

# TO CONNECT IS TO UNDERSTAND MATHEMATICS 4

**Nephroid** as envelope of circles
Choose a point G on a circle. Draw a circle with center at G that is tangent to a fixed diameter of the circle. Trace this circle as you drag G around the circle. The envelope of the circles is a nephroid.

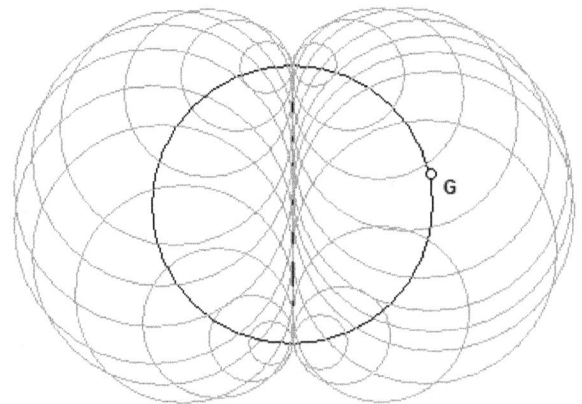

Figure 15


**References**
Boltianski, V. G. *La envolvente*. Moscow: MIR, 1977.
Jackiw, N. *The Geometer's Sketchpad 3.0* (Computer software). Berkely, CA: Key Curriculum Press.
Johnson, D. A. *Paper folding for the mathematics class*. Reston, VA: National Council of Teachers of Mathematics, 1995.
Lockwood, E. H. *A book of curves*. Cambridge: Cambridge University Press, 1976.
Yates, R. C. *Curves and their properties*. Reston, VA: National Council of Teachers of Mathematics, 1974.

# 5 FIBONACCI IN THE FOREST[5]

*Grade Level*: 4 - 6

*Mathematics Concepts / Skills*
Number sequences, number patterns, mental computation, connections with nature, similarity, spirals.

*Science Concepts / Processes*
Patterns in nature, growth, shape and function.

*Prerequisite skills*
Identification of number patterns. Similar shapes.

*Objectives*
Students will
1) Study mathematical patterns in a number sequence.
2) Relate the terms of the mathematical sequence to patterns in nature.

*Rationale*
     Content background
Mathematics. Leonardo de Pisa, also called Fibonacci, studied the sequence whose first terms are 1, 1, 2, 3, 5, 8 (Struik, 1969). Fibonacci encountered this sequence when studying a problem of how a population of rabbits grows (see *Extensions*). The terms of this sequence, after the first two, can be obtained as the sum of the two previous terms. Rectangles whose sides are

---

[5] Flores, A. and Guest, A. (1997). Fibonacci in the forest. *School Science and Mathematics, 97,* 388-392. Used by permission of Wiley and School Science and Mathematics Association.

in a ratio that is close to the ratio of successive terms of Fibonacci sequence, such as 5 to 3, 8 to 5, have been used frequently by artists and architects because of their beauty (Huntley, 1970).

<u>Science</u>. Organisms often develop so that parts grow preserving the same shape, for example, scales in pine cones, or pineapples. When new parts are added to fit with the existing, it is common that they appear "where there is most space". These two principles often originate the formation of spirals in living organisms. For example, in the following diagram, formed by squares of increasing size we can see that they naturally form two sets of spirals.

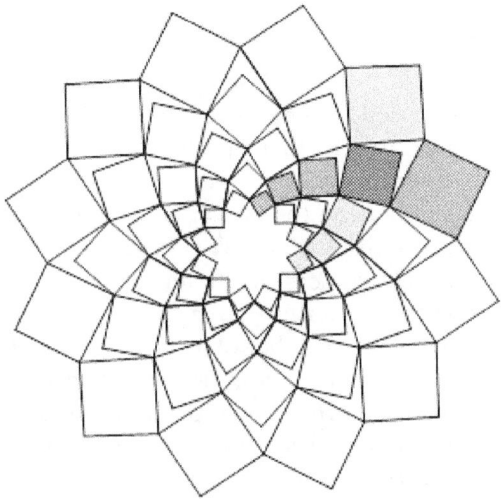

Figure 1. Growing squares form two sets of spirals.

A fact that is surprising for students is that the terms of the Fibonacci sequence appear frequently in these kinds of spirals in nature.

<u>Interrelationship of content</u>:
In this lesson children get actively involved in establishing connections between patterns in mathematics and nature. They explore how the Fibonacci sequence occurs in nature by collecting objects in the forest and analyzing them. The outdoor activity provides an opportunity for children to see that nature can be interpreted in mathematical terms. Students are always told that math is everywhere but this is hard for them to understand because numbers in nature are not always pointed out to them. Through this exercise the children get to see that patterns are part of nature and get a whole new outlook on math that makes it fun and interesting. Students see the significance of the Fibonacci sequence, as well as why things in nature grow

and live the way that they do. When they are in nature they will look for special patterns with numbers that will help them to appreciate their surroundings more.

Research background
Connections of mathematics with other fields are emphasized across all grade levels in the *Curriculum and Evaluation Standards for School Mathematics* (National Council of Teachers of Mathematics, 1989). *The National Science Education Standards* (National Research Council, 1996) emphasize unifying concepts and models. Mathematical models such as the number sequences and spirals discovered in this activity can help students see unity in a vast number of phenomena in nature.

*Lesson outline*
Before going outside to collect the pine cones, it is important that children explore the sequence from a mathematical point of view. The teacher can build towers made of unifix cubes whose heights are the first terms of the sequence; write the numbers on the board; let children guess the next term; teacher adds more term so that children verify their guesses; children describe properties of the sequence, such as that (after the first two terms) each term is the sum of the two previous terms. Then children are directed to go to the forest and collect pine cones, or other objects such as sunflowers. After the children have collected the pine cones, they gather together and are guided by the teacher to observe patterns in their objects. It is convenient to do the two discussions, the one that precedes the collecting and the one that follows it in a place where the teacher can show the sequence on the board or on paper, and where children are not distracted by the surroundings. If pine cones are not available, spiral patterns related to the Fibonacci sequence can also be found in pineapples, sunflowers, and daisies.

**Lesson**

*Purpose:* In this lesson, students will learn the Fibonacci sequence and become familiar with the concepts and patterns that occur in the sequence.
They will connect their knowledge of the Fibonacci sequence to objects in nature such as pine cones that show spiral patterns.

*Time:* The activity described here takes about 50 minutes with 4th and 5th graders. Exploring the Fibonacci sequence 15 -20 minutes. Recollection of pine cones or other objects in the forest 10 - 15 minutes. Discovery and discussion of patterns in objects 15 -20 minutes.

# TO CONNECT IS TO UNDERSTAND MATHEMATICS 4

*Materials for each student*
Unifix, Multilink, or wooden cubes
grocery bags
pine cones (or pineapples, or daisies or sunflowers that show a clear spiral)
paper and pencil
calculator

*Procedure*
*Discussion before the collection of pine cones.* The Fibonacci sequence is introduced by displaying towers of cubes that have the heights of the first four or five terms of the Fibonacci sequence (fig. 2). After students have observed the towers, ask how high will the next tower be. Why do you think so? Encourage use of mathematical language such as "sum," "adding," "contiguous."

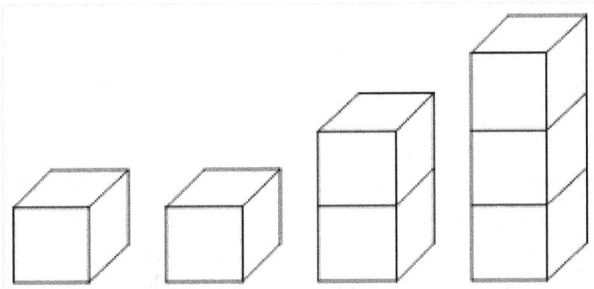

Figure 2. Towers of Fibonacci numbers.

After the concrete display, place the name of the sequence and put the first 4 numbers on the board: 1, 1, 2, 3. Ask the students if they recognize any patterns. Some children will recognize that there are two ones, some will say that each number has a comma between it, or that there is a difference of 1 between each number. Let other students verify what they describe. Next add another number to the sequence. The numbers 1, 1, 2, 3, 5 will be on the board. Ask again, "What types of patterns do you see?" Children begin to notice a pattern in the numbers related to addition, they will say..."Oh, 1 + 1 = 2, and 2 + 3 = 5." Sometimes they do not notice that 1 + 2 = 3, so make sure that it is pointed out so the children see it as part of the pattern that is starting to occur. Ask "What number comes next in the pattern?" Some of the children have now recognized an addition pattern and they will give the answer 8, and they will answer with "because 3 + 5 = 8, I added the 3 and the 5 to get 8." "Why did you choose to add 3 and 5?" "Because we added 2 + 3 to get 5 so to get the next number we have to add 3 + 5."
When students have a firm concept of the pattern, continue with the pattern using the adding process up into the larger numbers 1, 1, 2, 3, 5, 8, 13, 21, 34, 55, 89, 144, 233. To get these numbers keep asking the question, "How do

we get the next number in the sequence?" Ask students to brainstorm different strategies, write down the next number in the sequence and explain their strategy on how they got that number. The children may choose numbers like 21 + 34 = 55, which they may obtain by adding mentally 20 + 30 = 50 and 4 + 1 = 5, to get the answer 55.

Now that the children have taken the sequence up to a considerable amount of numbers, the teacher can hand out calculators and further ask, "Can you find any other special patterns that are occurring in the sequence besides addition?" The children will give different answers. If they are having a hard time, prompt them by giving the hint of looking at the numbers that are in the tens, hundreds and also the ones place. For example, a pattern found is that the digits 1, 2, 3, 5, 8 keep repeating along through the pattern, not only in the units place. Using this knowledge, the children may better understand which number will come next in the sequence.

*The walk in the forest.* Once the children have identified the Fibonacci sequence and understand the patterns and mechanics of the sequence, they are ready to break up and go on a little nature walk to choose special things that are found in nature. Instruct them to make sure that when they are looking for their nature products that they pick up some pine cones and flowers or other things they find interesting. This will provide some variety in nature's products. Instruct the students not to disturb the forest floor more than necessary. Also, do not take the last pine cone or flower in a group. This is a wise, Earth-friendly rule, taught by Native Americans.

*Discussion after.* Once the students have returned, have them all gather in a circle and have them pull out a pine cone if they found one. If they didn't find one, the teacher can give them one. Have each child place the pine cone between his or her thumb and pointer finger holding it upright so that it is almost on an axis. Model how to hold the pine cone and slowly turn it in their fingers. While they are doing this, ask them if they see anything special about the pine cone. Some students will answer that they see a spiral. Make sure that all the children have identified the spiral pattern formed by the cone scales (fig. 3). Let the children play with the pine cones until they all can see the spiral motion that occurs.

When everyone sees the spiral, model how to place their pointer finger on the pine cone in a specific place, turn the pine cone so that they can count all the spiral lines that occur in the spiral pattern. Ask them what number did they counted to. Some of the answers will be "I got the number 5," "I got the number 8" (if the activity is done with a pineapple some of them will get 8 and some 13; if using a sunflower, they may count 21 or 34 spirals).

Figure 3. Spirals in a pine cone.

Using the different answers, ask the students where they have seen these numbers before. They should recognize the numbers as terms of the Fibonacci sequence. Then have the students put the cone between their pointer finger and thumb again and rotate the cone in the opposite direction. They should see a second set of spirals, going the other way. Let them count the number of lines that make these spirals. The children that gave an answer of 8 the first time will give the answer 5 this time and the ones that gave 5 will get 8, depending on how they rotated the cone. Once again ask where the children have seen these numbers before and they will say the sequence. In each case the number of spiral lines for each cone, going one way is a term of the sequence, and the number going the other way is the next term. Point to the fact that the pine cone is composed by parts that have the same shape but that grow bigger in shape. The teacher can show at this point other examples where Fibonacci numbers and spirals appear in nature such as a pineapple, the center of a sunflower, etc.

Once the students have identified the sequence with something in nature and they have seen a connection, the teacher may guide the students to explore why the numbers of the sequence are on the pine cone or why the pine cone makes a spiral. The teacher may point out that a pine cone needs to grow but its parts stay the same shape. Teacher can point out that many things in nature grow conserving the same shape. For example, a seashell has a small mollusk living in it and when the animal grows the shell has to grow with it, but the chamber does not change shape. One way nature has to accomplish this is with a spiral. The teacher may share the following diagram, where the sides of the squares are given by Fibonacci numbers (see figure 4). The pattern is constructed by placing two unit squares sharing a side. Then a square with side two is adjoined to form a 2 by 3 rectangle. Then a square of side three is added to form a 3 by 5 rectangle, and so on. Once the discussion is over, the students do a reflection and write in their journals on the connection they just made between mathematics and nature and how they feel about it.

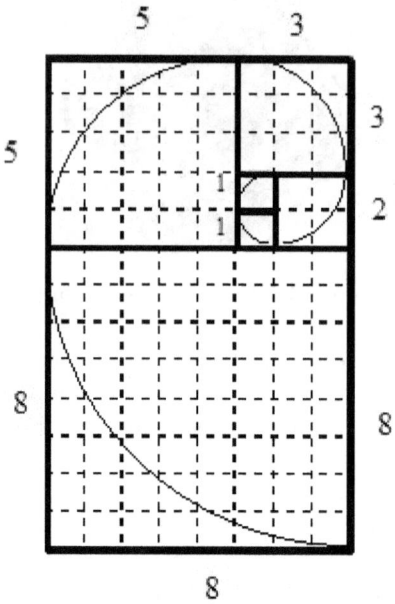

Figure 4. Fibonacci numbers and spiral.

*Evaluation*

The teacher interacts with students as they are doing the activity. Focus their attention on certain aspects, such as similarity of the scales of the cone. Students can describe how the sequence is constructed to their neighbors; show each other the spirals on the cone; drawing the pine cone, showing two sets of spirals; numbering the spirals in a different kind of cone, pineapple or daisy. Students can write in their journals about the most important ideas they learned in the activity, or about their feelings with respect the relation of math and nature. Reading what students wrote in their journals can be very informative.

*Extensions*

Students can also explore the relation of Fibonacci numbers with other aspects of plant growth, for example the distribution of leaves around the stem in some plants (Kappraff 1991). In some plants, if a helix is drawn to pass through each leaf until it reaches a leaf that is vertically above the first one, the number of turns of the helix and the number of leafs will be Fibonacci numbers (see table 1, taken from Huntley, 1970)

Table 1 distribution of leaves in plants

| Plant | number of turns of helix | number of leaves |
|---|---|---|
| Common grass | 1 | 2 |
| Sedges | 1 | 3 |
| Fruit trees (apple) | 2 | 5 |
| Plantains | 3 | 8 |
| Leeks | 5 | 13 |

Students can also count the number of petals in flowers. Fibonacci numbers appear in common flowers (Huntley, 1970; see also the poster by Garland, which illustrates some flowers, as well as the spiral patterns described above).

| | |
|---|---|
| Iris | 3 petals |
| Primrose | 5 petals |
| Ragwort | 13 petals |
| Daisy | 34 petals |

Fibonacci numbers also appear with animals. Students can study the rabbit problem that originated the sequence: A man has a pair of rabbits. These rabbits breed every month one other pair and begin to breed in the second month after their birth. How many pairs can be bred in one year? (Struik, 1969). Fibonacci numbers also describe the number of ancestors of the drone (Huntley, 1970). Students may also explore the different kinds of spirals on seashells.

**References and Resources**
Garland, T. H. *Fibonacci numbers in nature* (poster). Dale Seymour.
Huntley, H. E. (1970). *The divine proportion.* New York, NY: Dover.
Kappraff, J. (1991). *Connections: The geometric bridge between art and science.* New York, NY: McGraw Hill.
National Council of Teachers of Mathematics. (1989). *Curriculum and Evaluation Standards for School Mathematics.* Reston, VA: Author.
National Research Council. (1996). *National Science Education Standards.* Washington, DC: National Academy Press.
Struik, D. J. (Ed.) (1969). *A sourcebook in mathematics, 1200 - 1800.* Cambridge, MA: Harvard University Press.

# 6 SÍ SE PUEDE. IT CAN BE DONE: QUALITY MATHEMATICS IN MORE THAN ONE LANGUAGE[6]

Spanish and Native American languages have been spoken for centuries in many regions of the United States. Successive waves of immigrants have brought additional languages. In many regions, speaking multiple languages goes back several generations; in others, it is a recent phenomenon. With today's international competition and global economy, more and more people in this country realize that speaking more than one language is an asset. Of course, learning two languages requires additional effort, but as Leopold (1949, p. 188) states, "education does not make life easier, but better and richer. Few would condemn education for this reason. Bilingualism should be seen in the same light."

**Speakers of Other Languages and Their School Experiences**

Some adults remember that when they were children they were physically punished in school if they spoke their home language. Although there has certainly been a change, there is still a long way to go. Many schools equate limited proficiency in English with limited proficiency in academics in general. Only a few programs for gifted and talented students are offered in languages other than English. Many school districts do not try to teach in any language other than English. They argue that several languages are used

---

[6] Flores, A. (1997). Sí se puede. It can be done: Quality mathematics in more than one language. In J. Tentracosta (Ed.), *Multicultural and gender equity in the mathematics classroom* (p. 81-91). Reston, VA: National Council of Teachers of Mathematics. Copyright National Council of Teachers of Mathematics. Used by permission.

in the schools and that they obviously cannot offer such a variety. This argument is offered even in districts where one of those languages may be spoken by the vast majority of the children. Sometimes students who are not proficient in English are not enrolled in necessary content courses, even though language development is attained better when language is used in context. In many classrooms, students are prevented from discussing and explaining to one another in their own language "because they have to learn English." In some schools a disproportionate number of children who do not speak English well are labeled as learning disabled and put into special classrooms because of inadequate testing instruments (Spicker and McLeskey 1981). Tracking and other school practices limit the opportunities of language-minority students to learn mathematics (Oakes, 1990). In many classrooms, bilingual instruction is left to teacher's aides, rather than certified teachers.

## Underrepresentation in mathematics

Students who speak other languages remember that there was overt discrimination in the past. In the present they still experience in many places, subtle but pervasive, discrimination even by well-meaning individuals. Therefore, it is not surprising that historically in the United States some language minority groups have been underrepresented in mathematics. There are three related areas of underrepresentation: achievement in mathematics courses, enrollment in mathematics courses, and participation in mathematically related careers (Valverde, 1984). Low achievement can prevent enrollment in higher courses, and lack of enrollment in mathematics courses can also prevent participation in mathematically related careers. At the same time, low participation in those careers causes a lack of role models for future generations.

## A shift in paradigms

In the past, the predominant paradigms in research and practice when dealing with the low achievement of language minority students have been deficit models. "Such models concentrate on pinpointing and describing what students do not know, what experiences they presumably do not have, or what language and behavior differences they possess that result in a mismatch with the norms of the school" (Khisty, 1995, p. 279). Rather than focusing on changing the student and/or the family, rather than focusing on remedying the assumed deficiencies, we can look at the problem in a very different way. What we need to change is school practice in order to address the needs of language minority students. The recommendations given below are already practiced by many teachers in real schools. When a recommendation is presented, it will be illustrated with its implementation in the classrooms of the teachers who are described below.

**Three teachers**
Julia teaches a multi-age, bilingual classroom, grades 2 and 3, in an urban elementary school where 90% of the students qualify for free lunch. Students in her school come from various backgrounds, but in Julia's classroom almost all children have some Spanish in their lives. Julia first studied Spanish in high school and college. However, she acquired Spanish mostly by living in Honduras, by traveling with students' families in Mexico, and by studying in Spain. She occasionally asks students how to say a word in Spanish. Students help her in the same natural way that she helps them with English words.

Renee teaches a combined 4th and 5th grade class in the same school as Julia. Spanish is Renee's second language and her fluency was improved by teaching in Guatemala.

Julia and Renee have created classroom environments that are full of life. The walls in Renee's classroom are covered with students' work and art, newspaper news, and art posters (Picasso, Frida Kahlo). There is a turtle, hamster, fish, and green lizards living in the classroom. Books in English and Spanish on a wide range of topics fill the shelves.

Isabel, whose first language is Spanish, teaches mathematics to bilingual students in an urban High School in Phoenix where students come from a variety of ethnic backgrounds. When she entered school at age 5, the teacher told her family that they should speak to her only in English. She developed fluency in English but her Spanish was frozen and locked for many years. A few years ago she started working with bilingual students and, as she helps the students develop their English, they in turn help her develop and unlock her Spanish. Isabel takes great care to help students pronounce English words like "eighths" correctly. Unfortunately, as she told me, it is still common that people who have a foreign accent are deprecated.

The examples that are used to illustrate the recommendations were observed during visits to each of the teachers' classroom. Julia's class worked on arithmetic problems and geometry. Renee's studied empirical probability and proportions, and Isabel's 9th grade worked on measurement and algebraic expressions.

## RECOMMENDATIONS AND CLASSROOM EXAMPLES

**Have high expectations**
Students with talent in mathematics can be found in any ethnic group. Language minority students should be present among future scientists, mathematicians, and engineers in the same proportion as they are in the whole population. Teachers should have the same expectations for all groups to continue advanced studies and learn the necessary mathematics. Schools should provide information about the educational requirements for careers

that make use of advanced mathematics. Students should be encouraged, guided, and prepared to enroll in mathematics courses beyond the minimum requirements.

Gifted and talented students are equally represented in all groups. However, gifted and talented potential will not develop unless properly nurtured and encouraged. Students who speak languages other than English should be expected to participate in activities that nurture their gifted potential in the same proportion as other groups. Mathematics activities in other languages must be provided.

The school where Isabel teaches used to list grades from bilingual students separately from the other students, even though the courses were exactly the same. The implicit message was that the expectations were not the same for them. Isabel convinced the administration that there should not be a separation. Now the message is that expectations are the same.

## Provide the same high quality education for students who speak other languages

Higher order thinking is a basic skill for all students (Chancellor, 1991). Making conjectures and testing them, making inferences, giving convincing arguments, and generalizing are skills that all students need to practice and develop.

Renee's class conducts a probability experiment in which students simulate a race by tossing two dice and naming the sum. The first sum to appear nine times is the winner. Before beginning, Renee asks students to predict which sum will be the winner and to mark their favorite. Students work in small groups, tossing the dice and marking the outcomes. Some students call the numbers in Spanish. All students participate eagerly in the activity when they have to conjecture and experiment, despite the wide range of skills.

We need to attract language minority students to mathematics at an early age, and retain them. This can be done best by developing higher-order thinking skills, and using teaching and learning approaches that are sensitive to their learning needs, including the language. In one session Julia gives each child different three-dimensional figures: cylinder, sphere, square pyramid, cube, hexagonal prism, cone, etc. She does not tell the students anything about the shapes and asks "¿Cómo puedes describirlo a un amigo?" [How can you describe it to a friend?]. Children describe their shapes in English or Spanish, while the other students listen. They use informal language, appropriate for their level of development in geometry, such as "un pico y cuatro esquinas" [one point and four corners]. They use two-dimensional shapes to describe the faces and cross sections of their solids as seen from different angles. They clearly see the difference between two- and three-dimensional shapes. For example, a child, when describing the cone, says that it looks like a triangle, but that the triangle cannot hold ice cream, but the cone can.

Students who speak other languages, as any other group, vary widely in their learning styles. For example, some students feel more comfortable learning in an environment where small group learning encourages cooperation rather than in an environment that stresses competition. In Julia's class, pairs of children play "circles and stars". They throw one die and draw that number of circles; then they throw the die again and draw that number of stars inside the circles. Students then find the total number of stars (by counting). Some write an expression like $2 \times 1 = 2$ for two circles with one star each, others simply write the answer. Students check each other's work. In the meantime, two girls play a form of Nim where they take turns removing one or two objects from a collection, and the person who takes the last one loses. One of the girls is able to predict that she will win a few moves in advance, and she wins most of the time. The other girl does not give up, trying different approaches. She seems to enjoy the game even though she has not figured out a winning strategy.

The drastic improvement in performance in mathematics of bilingual students has been noticed by Isabel's school. Now she is invited to share with other teachers what she does. After a demonstration of how to teach bilingual students one teacher said to her: "But you didn't do anything *special.* That was just good teaching." Isabel answered: "That is the whole point."

## Provide multiple points of entrance to higher-order thinking in mathematics

Many times, language minority students get trapped in a vicious cycle of "remedial" courses. They cannot take part in more interesting and challenging courses in mathematics because they lack "basic" skills in English or in mathematics. The acquisition of a language takes several years, the acquisition of mathematical concepts too. Concept learning should not be put on hold while students develop proficiency in English.

In Isabel's school, recent immigrants do not have to wait to take mathematics even if they don't speak English. The school provides bilingual mathematics courses for Spanish speakers, and English as a Second Language mathematics sections for students who speak other languages. Isabel also gives her students opportunities to acquire basic mathematics skills in a meaningful context. Her students, recent immigrants, grew up using the metric system, and they need to learn the English system. Students use measuring tapes to measure their wrist, head, arm, and height in inches and fractions of an inch. They work in pairs and help each other; they are on task, use a variety of methods to measure, and they compare measurements with each other. Isabel does not keep them on hold to learn more advanced material while these basic skills are acquired. In the same session they work with algebraic expressions.

If students are not proficient with elaborate paper and pencil computational skills, they can use a calculator and focus on how to solve a problem. They can estimate and see whether the answer is reasonable and don't need to be bogged down by the computations. Isabel hands out calculators to every student for an activity involving algebraic expressions. She writes several at the board with instructions like "Solve and explain", "Solve for $x$ / explain", "Simplify". She encourages students to work in small groups or in pairs. Students work together and explain to each other, mostly in Spanish, the steps that are necessary to do the exercise. The teacher walks around the room, asking and answering questions. When students have completed out the problems, they explain the exercises at the board, by writing the steps and briefly jotting an explanation for each step. When one student hesitates when adding (-15) + 4, the teacher reminds him that he can use a calculator, and helps him to understand why the answer is correct.

Another tool that provides access to important concepts for students who lack skills in algebraic symbolic manipulation is the graphing calculator. Using the calculator, students can connect different representations – graphs, tables, and equations – of functions. With today's technology, students can have access to probability, statistics, and geometry. The necessary computational skills can then be developed in a meaningful context.

Manipulative materials and concrete representations can provide another point of entry. In this context, children who have difficulties with traditional instruction can have success and show their talent. Higher order thinking skills such as problem solving, reasoning, conjecturing, experimenting, evaluating, and communicating can be developed in the context of using concrete materials.

Julia has posed a problem of how many tires are needed for 4 bicycles and three tricycles (Burns, 1988) which is an opportunity to deal with multistep problems. A child solves the problem by making a drawing and recording partial results (figure 1). Then he adds 8 + 9 mentally to obtain the answer.

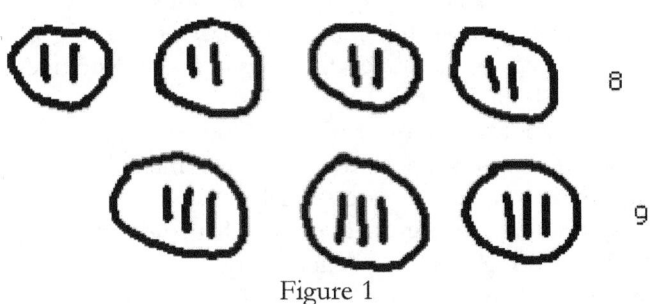

Figure 1

Isabel lets her students use whatever materials are appropriate. For the

Pythagorean Theorem, the students used cardboard pieces to form a puzzle (Hall, 1974). She spent two sessions guiding the students and helping them figure out the relations among the pieces. She used a variety of questions, rather than telling them the results. On the departmental test, all students in her classroom answered the questions on this theorem correctly.

## Do not short-change the students, emphasize all aspects of mathematical discourse

Manipulative materials are often used in mathematics to circumvent communication problems between students with limited English proficiency and teachers with no proficiency in the student's language. However, it is important to realize that activities with concrete materials are not enough. Students learn not only because of the experiences with concrete materials, but mostly because of their reflection on those experiences (Pirie, 1988). Talking about the experience provides the opportunity to reflect, as Cazden (1986) highlights. Students should have the opportunity to discuss their mathematical activities and findings in their first language and in English. Therefore, we need to foster the development of mathematical vocabulary in both languages.

Julia is working mostly with second graders in this activity. She writes addition problems on the blackboard. Children compute mentally (and with their fingers) using their own strategies. They then take turns giving their answers and explaining their methods. For a problem like $3 + 8$, some students use counting-on aloud (in Spanish), some use fingers, others count in their heads, some count from three, a little girl counts-on from eight, another student adds "$8 + 2 = 10$, and one is 11". Children occasionally disagree on an answer and use mathematical arguments to convince themselves and other students. One student has doubts about the result of $7 + 7$, and says "seven" and then counts by keeping track with the fingers "eight, nine, ten, eleven, twelve, thirteen, fourteen" to convince himself. On the problem $3 + 2 + 8 + 7$, one student adds $3 + 7$ and $2 + 8$, and then adds $10 + 10$ to get the final answer. A girl mentally adds $8 + 7 = 15$, and $3 + 2 = 5$, and then she counts on five more starting from 15 to get the answer. This part of the session is a good example that children can reinvent arithmetic (Kamii, 1985; 1989).

There is also another reason that makes communication very important in mathematics. Communication of the internal, mental representations of concepts and their connections is essential to see whether the student has attained understanding. Research has shown that students can reach correct answers by using strategies that reflect very little understanding of the problem (Sowder, 1988). It is important that students explain how they arrived at the answer in the language with which they feel most comfortable. When a child tries to explain his counting strategy in English, Julia realizes that at this point his counting in English is by rote, but that he can count meaningfully in Spanish. She encourages him: "Si quieres contar en español..." [If you want, count in Spanish].

In Renee's class students work in small groups conducting probability experiments. Each group conducts a slightly different experiment (they use different objects: plastic cup, styrofoam cup, a bottle lid, a cardboard cylinder). In each case, they drop objects and record how many times they land in a given position (50 tosses). There are three possible outcomes, each marked with a letter A, B, or C. In the meantime, the teacher walks around the classroom and interacts with different teams, clarifying, probing, questioning. Students engage in whole class discussion, using both Spanish and English, on how to compute the experimental probability from the ratio of outcomes in 50 tosses. Renee focuses their attention on the fact that outcomes are not equally likely due to the shape of the objects. Renee guides students to use proportional reasoning to figure out the empirical probability given the outcomes in 50 tosses. After the experiments are over, the whole class forms a circle to share comments about the activities. The teacher probes their thinking with questions like: "¿Por qué piensan que cayó C más?" [Why do you think C came out more times?]. Students use their own language to explain why. For the cylinder, one student said that it was more likely to fall on the side. Her explanation of why was "Es más fuerte lo del centro que la orilla" [In the middle it is stronger than on the rim]. The teacher also lets them express their feelings about the activity. One student expressed how exciting the race was: "En los juegos era mucha emoción" [In the games it was very thrilling].

Isabel uses a number line to introduce expressions like "$x$ is equal to 12," and the symbolic representation "$x = 12$." The expression "$x$ is less than 6," is also represented on the number line. She provides many opportunities for all students to develop their communications skills in mathematics. Isabel gives them ample time, asks many guiding questions, and lets students formulate their own description of the relations between the numbers. Students use their own words, sometimes using informal expressions like "una cadena de números" [a chain of numbers] to describe intervals.

Mathematical discourse in small groups also promotes cooperation among students and encourages them to take an active role. Renee encourages students to discuss the outcome of the probability experiment: "Hablen entre ustedes. ¿Por qué pasó así? ¿Qué piensan? Y después escriben." [Talk among yourselves. Why did it happen this way? What do you think? And then write]."

## Use fair and meaningful assessment and testing procedures

A person who is not proficient in English will be at a disadvantage if the mathematics test is in English. However, a test written in the language of the student may still not be fair, if students have not been taught the technical vocabulary used in mathematics, if students did not have the opportunity to learn to write and read in their language, or if the test is a poor translation.

Isabel's students have learned to read in Spanish in their countries of origin, but their level of English proficiency would not allow them at this point to show what they know in mathematics if the test were in English. She provides a Spanish translation of the departmental test. Assessment of mathematics in other languages should include measures of high level conceptual knowledge, high level procedural knowledge, and higher order thinking skills.

**Develop language proficiency related to content learning**

Proficiency in a second language is best accomplished when proficiency in the primary language continues to develop. Language development should be related to content (Crandall, Spanos, Christian, Simich-Dudgeon, & Willetts, 1987), and mathematics should not be an exception. At the same time, the mathematics teacher should also be aware that some of the academic language used in materials and discussions in the mathematics class may be especially difficult (Cuevas, 1984). The mathematics teacher should also be a teacher of the language needed to learn concepts and skills in mathematics. A consistent finding reported by Valverde (1984) is that bilingual students do better in mathematics when taught bilingually than monolingual English-speaking Hispanic students, or students with a limited proficiency in English when taught monolingually.

Julia, Renee, and Isabel use both languages to conduct their lessons. Students have the freedom to use either language, but the teachers are very careful that mathematical terms are learned in both. In her lesson on probability Renee writes on the blackboard the terms: proportion, proporción, probability, probabilidad. Students use their own words to make sense of the new terms, and then use the new terms in both languages.

Schools should try to provide high-quality mathematics materials and books in languages other than English. There are Spanish versions of exemplary materials such as the *Estándares curriculares y de evaluación para la educación matemática* (NCTM, 1991a), and *Matemática para la Familia* (Stenmark, Thompson & Cossey, 1987). However, the lack of enough mathematics materials for bilingual students is still a problem.

**Be proud of the cultural heritage**

The teaching and learning of mathematics is certainly not a "culture free" endeavor, and teachers and textbooks should provide examples of past and present contributions to mathematics, science, engineering and other aspects of culture with mathematical elements from people representing many languages. The curriculum should include examples from the rest of the world as well as from people in the United States whose first language is not English. Language is an important part of the culture. For language minority students, learning the language at school will improve the proficiency of the language spoken at home and help preserve their cultural heritage.

In Isabel's class, many students have learned algorithms in their countries of origin that are different from the algorithms commonly taught in this country. She helps students to compare and understand the differences. She uses whatever algorithm is more convenient for a particular situation. She also learns and presents to her class alternative algorithms from books published in the countries of origin of her students.

## What if the teacher is not multilingual?

In some schools it would be impossible for any teacher to speak all the languages represented. Nevertheless, the teacher can help children to continue to develop proficiency in their first language at the same time that they develop proficiency in English. The basic premises are respect and encouragement. Students should feel free and proud to express their mathematical thinking in their first language.

In my mathematics methods course, I have students that speak Apache or Navajo – languages I do not understand. The following explanation from a portfolio entry reflects how important it is for students to be encouraged to use their language even if the teacher does not understand it.

*dii le k i dinldoo Problem Solving*
I included this because this was when I first experienced with writing in my Apache language. I have read and seen the Apache language written in short articles and in the dictionary. However, I had never attempted to write until I came to this class. Now that I know its relevancy, I plan to stick with writing in Apache and reading. I will also incorporate the Apache lessons as a teacher.

Schools that have many languages represented also have a wealth of resources available that they can tap. Students can help other students. Students working in small groups that speak the same language can use it as well as English in their discussions. Students with a greater proficiency in both languages can serve as mediators between the teacher and the other students.
Students who prefer to write their reflections in mathematics in their mother language can receive reinforcement and feedback from parents and other community members.

School districts have successfully used different strategies: magnet schools for different languages; two-way immersion programs where everybody learns two languages. Some deal with one of the languages first, find what works for them and then address the other languages. Schools can group students so that the number of languages in a single classroom is not too high.

## CONCLUSION

In the classrooms described above students show many of the traits stated as goals in NCTM's *Curriculum and Evaluation Standards* (1989): they value mathematics, they are confident in their ability to do mathematics, they are mathematical problem solvers, they communicate mathematically, and they reason mathematically. Teachers show many of the traits described in the *Professional Standards for Teaching Mathematics* (1991b). As in the case of Isabel, what they practice is "just good teaching" using two languages. In short, recommendations on how language minority students can best learn mathematics and how they should be taught do not differ significantly from what is best for other groups: students and their teachers should be proud of who they are, and should use their cultural heritage, including their language, to their advantage. Teachers should have high expectations, and provide high quality instruction and learning opportunities for their students. Teachers should use approaches to mathematics and teaching styles that fit the needs of their students. Assessment and testing should be unbiased and should include higher order thinking skills.

Coordinated and continuous actions are needed to achieve this equity agenda. It is not an easy task; it does not make life easier. However, there are many concrete examples of teachers and schools that in their everyday practice implement the pedagogical, curricular, and policy recommendations described above to provide quality mathematics in more than one language and make students' lives richer and better. Sí se puede.

## REFERENCES

Burns, Marilyn, and Bonnie Tank. *A Collection of Math Lessons from grades 1 through 3*. White Plains, N.Y.: Cuisenaire, 1988.

Cazden, Courtney. B. "Classroom discourse". In *Handbook of research on teaching*, 3rd ed., edited by Merlin C. Wittrock, pp. 432-463. New York, N.Y.: Macmillan, 1986.

Chancellor, Dinah. "Higher-order thinking: a basic skill for everyone." *Arithmetic Teacher* 38, (February 1991): 48 - 50.

Crandall, Jo Ann, G. Spanos, D. Christian, Carmen Simich-Dudgeon, and Karen Willetts. *Integrating language and content instruction for language minority students*. Silver Spring, Maryland: National Clearinghouse for Bilingual Education, 1987.

Cuevas, Gilbert J. "Mathematics learning in English as a second language." *Journal for Research in Mathematics Education* 15 (1984): 134-144.

Hakuta, Kenji. *Mirror of language: The debate on bilingualism*. New York, N.Y.:

Basic Books, 1986.

Hall, G. D. "A Pythagorean puzzle." In *Teacher made aids for elementary school mathematics*, edited by S. E. Smith and C. A. Backman, pp. 142 - 145. Reston, Va.: National Council of Teachers of Mathematics, 1974.

Hiebert, James, and Thomas P. Carpenter. "Learning and teaching with understanding." In *Handbook of research on mathematics teaching and learning*, edited by Douglas A. Grouws, pp. 65-97. New York, N.Y.: Macmillan, 1992.

Kamii, Constance. *Young children reinvent arithmetic.* New York, N.Y.: Teachers College Press, 1985.

Kamii, Constance. *Young children continue to reinvent arithmetic.* New York, N.Y.: Teachers College Press, 1989.

Khisty, Lena Licón. "Making inequality: Issues of language and meanings in mathematics teaching with Hispanic students." In *New directions for equity in mathematics education*, edited by Walter G. Secada, Elizabeth Fennema, and Lisa Byrd Adajian, pp. 279-297. New York, N.Y.: Cambridge University Press, 1995.

Leopold, Werner F. *Speech development of a bilingual child*, Vol. 3. Evanston, Ill.: Northwestern University Press, 1949.

National Council of Teachers of Mathematics. *Curriculum and evaluation standards for school mathematics.* Reston, Va.: The Council, 1989.

National Council of Teachers of Mathematics. *Estándares curriculares y de evaluación para la educación matemática.* Reston, Va.: The Council, 1991a.

National Council of Teachers of Mathematics. *Professional standards for teaching mathematics.* Reston, Va.: The Council, 1991b.

Oakes, Jeannie. *Multiplying inequalities: The effects of race, social class, and tracking on opportunities to learn mathematics and science.* Santa Monica, Calif.: The Rand Corp., 1990.

Pirie, Susan. "Understanding: Instrumental, relational, intuitive, constructed, formalised...? How can we know." *For the Learning of Mathematics* 8 (1988): 2-6.

Sowder, Larry K. "Children's solutions of story problems." *Journal of Mathematical Behavior* 7 (1988): 227-238.

Spicker, Howard H., and J. McLeskey. "Exceptional children in changing times." In *The mathematical education of exceptional children and youth*, edited by Vincent J. Glennon, pp. 1-22. Reston, Va.: National Council of Teachers of Mathematics, 1981.

Stenmark, Jean Kerr, Virginia Thompson, and Ruth Cossey. *Matemática para la familia.* Berkeley, Calif.: EQUALS, 1987.

Valverde, Leonard A. "Underachievement and underrepresentation of Hispanics in mathematics and mathematics related careers." *Journal for Research in Mathematics Education* 15 (1984): 123-133.

# 7 TIN-CAN ICE CREAM[7]

*Grade Level:* 5 - 9
Note: Actually, children of all ages enjoy making ice cream this way. With smaller children, the teacher can help more with the measuring, and the discussion can be kept on a less technical level. On the other end, students in a physics course can use the activity to illustrate concepts about heat.

*Mathematics Concepts/Skills*
Fractions, length (feet/meters), time (minutes), volume and capacity (cups and teaspoons/liters and milliliters).

*Science Concepts/Processes*
Melting point of water, melting point of ice and salt, change of phase of water (condensation and melting), energy transfer in the form of heat, temperature changes associated to change of phases, freezing mixture.

*Objectives*
Students will
    1) use measures of the English (also called customary) or of the metric system in a real-life context by using standard household measuring cups and spoons to
       a. accurately use full and partial units of standard measures;
       b. promote personal success using simple tools as opposed to commercially produced machines which should be operated by adults.
    2) use fractions of unit measures, and demonstrate the binary nature of

---

[7] Flores, A. & Perkins, I. (1996). Tin-can ice cream. *School Science and Mathematics, 96,* 46-49. Used by permission of Wiley and School Science and Mathematics Association.

standard divisions, that is, halves, quarters, and eighths, and/or as the decimal divisions of metric units of measure;
3) observe changes in the phase of water: ice to water; vapor to liquid to ice;
4) observe corresponding changes in temperature by noticing and sensing changes in the container as the activity is carried out.
5) understand the principles of ice cream making; and
6) make ice-cream and enjoy it!

*Rationale*

The lesson described uses students' active participation in the making of ice-cream. Students work in groups and cooperate to reach the goal. Students have an opportunity to practice fractions and use measurement instruments of the English or the metric system in a real-life activity. Students also observe changes in the phase of water and notice associated changes in the temperature. Students have the experience that when water is heated, it changes from solid (ice) to liquid, or from liquid to vapor. This activity provides the opportunity to see that when there is a phase change there is a transfer of heat, and a drop in the temperature can be observed.

This activity can be used to address the *Standards* (NCTM, 1989) on measurement that students should "make and use measurements in problem and everyday situations" (p. 51), and so that students can "understand the structure and use of systems of measurement" (p. 116). The activity can also be used to illustrate and discuss important concepts related to energy, such as energy transformation, energy conservation, and heat energy as recommended in the *Benchmarks for science literacy* (Project 2061, 1993).

*Background information*

Water can exist in three phases: ice, liquid, or vapor. When water changes from one phase to another, energy in the form of heat must be transferred. When the phase change is from a liquid phase to a gas phase or when the change is from a solid phase to a liquid phase, the water must absorb heat (see Halliday, Resnick, Walker, 1993). When water evaporates, the heat is absorbed from the surroundings. Thus, when water evaporates, the temperature around drops. This principle has been used widely in many cultures to cool houses in hot climates in the form of fountains in Spanish and Arabic patios, evaporative coolers in U. S. Southwest, lavish inner gardens in Yucatan, etc.. When ice melts, there is also a drop in the temperature in the surroundings. People in regions with cold winters notice that when the sunshine melts the ice and the snow, it does get *colder*.

People have also noticed that adding salt to water changes the freezing point of water. Pure water becomes ice at 32° F (0° C) at sea level. However, for saltwater the temperature needs to be lower to become ice. The water in the oceans freezes at 28° F (-2° C). In fact, Gabriel Fahrenheit used the lowest temperature he could get with ice, salt, and brine as the 0° on his scale. Therefore, if we add salt to (pure) ice it will start to melt if it is not too cold (remember that ice can be as warm as 32° F but as cold as -6° F or even colder!). Cities in different latitudes have different policies with respect to the use of salt to melt ice and snow on the streets. In Columbus, Ohio, winters are not too cold and salt is used to melt ice and snow. In Minneapolis, Minnesota, however, it gets so cold, that adding salt would not solve the problem. The reason is that salt would melt the snow, but if the temperature drops even more, the water (even with the salt) would get frozen into ice again.

There is another interesting phenomenon happening when we add salt to ice. If ice, salt, water, and vapor at 32° F (0° C) are together in a container that is thermally insulated, they will not be in equilibrium, and some ice will melt and dissolve some salt (see Zemansky & Dittman, 1981). This saturated solution will be too concentrated to be in equilibrium with ice, therefore some ice will melt, lowering the concentration of the solution, which in turn will then dissolve more salt. While this is going on, the temperature of the whole system automatically and spontaneously decreases until the temperature of about -6° F (-21° C) is reached. This system is known as a *freezing mixture*.

So, in this activity, when salt is added to the ice inside the can, the ice not only melts, but the temperature drops. Students will notice how quickly the temperature of the can drops, and how fast the outside of the can gets wet due to condensation of water in the atmosphere. In fact, a thin layer of ice is sometimes formed on the outside of the can in this activity. The idea of using two cans to make ice cream, and the recipe were taken from Thomas (1982). The can is rolled to ensure that all the ice cream ingredients mix and at the same time get in contact with the side of the can, which is in contact with the freezing mixture of ice and salt.

*Lesson outline*

Before the activity, students need to understand how they are going to make the ice cream, and discuss why it works. The session before, discuss different aspects that you want them to observe during the actual doing of the ice cream, such as dropping of temperature, condensation, melting, etc.. Giving instructions and carrying out the procedure to make ice cream takes about 45 minutes, so at least one session is needed just for that. The teacher hands out a copy of the recipe to everyone, forms teams (up to a total of 6 students per team). The teacher goes through the recipe at the board or overhead and models how the small can is put inside the bigger can, packed with ice and salt (see Figure 1), taped, and rolled back and forth. Student groups prepare their own cans, roll them back and forth, and eat the ice cream.

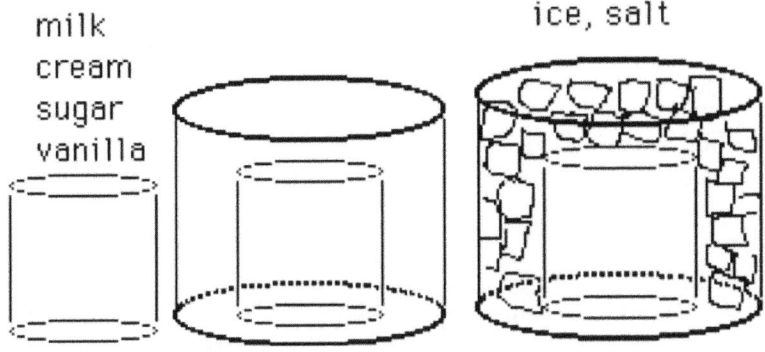

Figure 1. Packing the ice cream can in the freezing mixture

After the ice cream is done the teacher can wrap up the discussion, ask students to describe what they observed and listen to their explanations as to what happened and why.

**Lesson**

Purpose: In this lesson students practice fractions and measurement in the English system or the metric system as they make ice cream. They observe changes in the state of water and changes in temperature.

Time: Instructing the students on the procedure, making the ice cream: 45 minutes (students may need additional time to eat and clean up).
Another 20 minutes are needed before the activity to explain how they are going to make the ice cream and discuss why it works. The activity is followed by another discussion (about 15 minutes).

Materials/Supplies for each team (up to 6 students). (Alternative: use metric units)
one 3-pound coffee can with lid
one 1-pound coffee can with lid (actually the cans are 12 to 13 ounces)
ice (one 8-pound bag is enough for four teams)
1 cup rock salt
masking tape or duct tape (duct tape does not slip when it gets wet)
small towels
measuring cups (cups are usually marked in both metric and standard units)
measuring spoons (some teaspoons have the 15 ml and 5 ml equivalents)

cups, spoons, and napkins to eat the ice cream
tape or gauge to measure 6 feet (or make sure students can estimate 6 feet)

Ingredients for each team (Alternative: use metric units)
1 cup milk
1 cup whipping cream
1/2 cup sugar
1/2 teaspoon vanilla extract
(chocolate chips and sparkles can be added later if desired)

*Procedure*

Discussion before the activity. The process of ice cream making is described. Teacher explains what students will do, and emphasizes what might happen if lids are not tapped well, or if students sit too far apart to roll the can, etc.. The class discusses also why this method for making ice cream works. Teacher may want to focus students' attention on particular aspects of the activity: melting, condensation of water in the air, drop of temperature, etc.. (Teacher may also want to tell class in advance that since she/he will not be able to eat ice cream from every team, she/he will not eat ice cream from any team, so that students' feelings will not be hurt.)

Instructions for the activity. Teacher goes step-by-step with the whole class over the recipe, ingredients, materials and procedure of making ice cream. Teacher models rolling the can back and forth with another student. Teacher forms teams which are composed of one captain, one head cook, and shakers (up to six members per team).

The making of the ice cream. The captain and the head cook of the team measure all ingredients and put them in a 1-pound coffee can that has a plastic lid that fits tightly. They tape the lid with masking or duct tape. They put the can with ingredients inside the 3-pound can with crushed ice around smaller can. They pour 3/4 cup of rock salt evenly over the ice. They place lid on 3-pound can and tape it tightly. The other students take turns in pairs at two minute intervals to roll can back and forth on the floor. Students sit up to 6 feet apart from each other (that way it is easier to control the can, if they sit further away the can tends to slow down and the cream and milk mixture does not freeze as well, and students will get tired). They will roll the can for 10 minutes (in the beginning they may have a little problem making the can roll straight). Then the outer can is opened, and the can with the ingredients is removed. Students remove the lid making sure no salt water gets into the ice-cream mixture. They use a spoon to stir up the mixture and scrape the sides of the can (the part of the mixture that is in contact with the can will be frozen, and the mixture at the center may still be liquid). Replace lid and tape again. Teacher drains ice water from larger can. Students insert small can again, and they pack with more ice and salt. Students roll back and forth for five more minutes. This recipe makes enough ice-cream for all team-members.

Discussion after. The ice cream activity can be used as a wrap up for a unit on measurement using the English system. The students can also share their observations of what happened while they were making the ice cream, and share their explanations. Teacher can guide the discussion, and assess students' understanding through questioning.

*Evaluation*

Teacher interacts with students as they are doing the activity. Focus their attention on certain aspects, ask and answer questions. A clipboard with student's names may assist in evaluation during the activity.

Questions. Students' explanations can give us insight into their understanding. The answers used as examples are real students' answers.

*Did the can always roll straight to your partner? Did it sometimes go to the right or to the left? If so, why do you suppose that might have happened?* One student said that the can did not roll straight because of the lid. One end was slightly bigger than the other end and because of this the can was lopped-sided and could not roll straight. Another student said that the ice was melting inside the larger can and some of the ice which was still in chunks moved around in the large can causing the can to roll crooked. Some students said that the ingredients in the small can were sloshing around and that the ice as it melted sloshed around and the two sloshing liquids pulled the can in different directions.

*What do you suppose is happening to the milk, cream, sugar, and vanilla while the cans are rolling back and forth? Make a sketch to show what is happening to all the ingredients as they are rolled; do not forget to include the ice and rock salt. Be sure to label your drawing. Explain what is happening in your sketch.* A sketch can be drawn on the board. Students may use a lot of hand waving to describe the movement of the two liquids.

*Why is it important to* roll *the cans instead of just* pushing *them to your partner?* Help your students understand that the ingredients in the smaller can have to make contact with the "cold" in the surrounding ice in the larger can. If there is not enough contact the ingredients would not turn into ice cream. They would stay liquid or be mushy.

*Some teams see a thin layer of ice or frost after the can rolls around for a while. Where did the ice come from? Did the can leak?* Some students may have some previous knowledge about condensation. Some will be surprised to find that there is water in the air.

*After the can has been rolled for ten minutes, it is sometimes hard to tape the lid onto the can. Why is it hard to tape the lid to the can?* Most students say that it is all the ice on the can, and that when you touch it, it turns to water and it makes it hard for the tape to stick.

*Extensions*

Students can use this activity to practice measurement in different systems. They can change the recipe to use metric units, and use metric measurement instruments.

Students can also practice proportional thinking by changing the recipe for more people, or by changing the units. For example: *Suppose you only had a 100ml beaker and you decided to make the ice cream anyway. You decide to use 100 ml of milk and of ice cream. How much sugar would you need to make the recipe? How much vanilla would you need?* (Hint: 1 cup is about 250 ml and one teaspoon is about 5 ml.)

The activity can also be conducted outdoors using snow instead of ice. The discussion can include the snow used. *Where did all this snow come from?*

Students may also use a thermometer to measure the changes in temperature when pure ice is used, and contrast to changes in temperature when salt is added to the ice.

## References

Halliday, D., Resnick, R., & Walker, J. (1993). Fundamentals of physics. New York: Wiley.

National Council of Teachers of Mathematics. (1989). Curriculum and evaluation standards for school mathematics. Reston, VA: National Council of Teachers of Mathematics.

Project 2061, American Association for the Advancement of Science. (1993). Benchmarks for science literacy. New York: Oxford University Press.

Thomas, D. (1982). Today's tips for easy living. Tucson, AZ: HP Books, Fisher Publishing.

Zemansky, M. W., & Dittman, R. H. (1981). Heat and thermodynamics, sixth edition. New York: McGraw-Hill.

# 8 BILINGUAL LESSONS IN EARLY-GRADES GEOMETRY[8]

The activities described in this article were done by all children in 21 bilingual classrooms in grades K, 1, 2, and 3, in a laboratory, hands-on setting. The goal was to engage young children in creative activities in mathematics, to provide experiences suited for gifted bilingual children at an early age, as part of Project Excel (Pérez, 1991). The written and oral instructions for the activities, as well as the discussion was conducted in Spanish, since one of the objectives of the project is to develop higher-order thinking skills in both languages, and many times bilingual children are not provided in schools with the vocabulary and other tools to do mathematical discourse in languages different than English. The activities presented in this paper are in concordance with NCTM's Position Statement on Early Childhood Mathematics Education (1991), NCTM's recommendations for gifted students (1987), and NCTM's Standards (1989, 1991). Several of the components essential for programs for gifted (House, 1987) are incorporated. The mathematical activities are worthwhile mathematical tasks, were done in a propitious learning environment where students could develop skills in higher order-thinking skills, applications and problem solving, and communication, with activities that encourage creativity, through the use of learning resources. The informal, intuitive geometry activities present important ideas and concepts and their interrelations to children, using concrete representations, and at the same time give gifted children the opportunity to explore a mathematical topic in depth. The

---

[8] Flores, A. (1995). Bilingual lessons in early-grades geometry. *Teaching Children Mathematics*, *1*(7), 420 - 424. Copyright National Council of Teachers of Mathematics. Used by permission.

activities are challenging enough for gifted children, but also are within the reach of average students, who also benefit from higher-order thinking activities in mathematics (Chancellor, 1991). Children in kindergarten did the first five activities in a session. A minimum of technical terms were used (square, triangle), and informal language was used ("the small triangle"). The number of activities is increased for higher grades and also the activity is adjusted for each grade. More vocabulary is added, more precise description of the figures is given (for example, "a square has four equal sides and four equal angles"), and quantitative relations among the figures are made explicit (for example, "the area of the square is twice the area of the triangle"). Right isosceles triangles of two sizes, and a square cut out from cardboard were given to the children (see copy of attached materials).

**Activity 1**
Children take the cardboard square and cover the square in figure 1, to see that both squares are equal. Then they use the cardboard square and cover figure 2, to see that figure 2 is also a square and equal to the first.

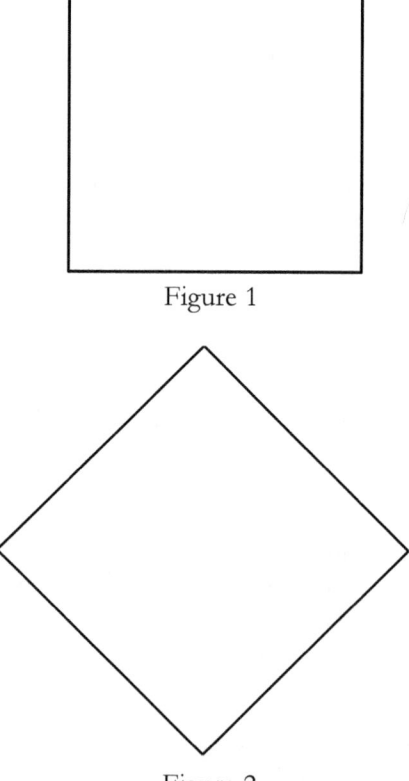

Figure 1

Figure 2

## Activity 2

Children take one of the small cardboard triangles, and they cover the triangle in fig. 3 with the cardboard triangle.

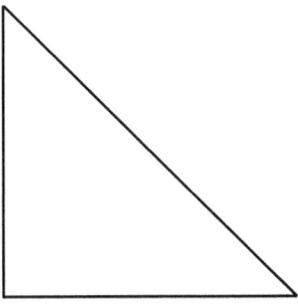

Figure 3

Next they cover with the cardboard triangle each of the triangles in Fig 4. to see that if they move the triangle, if they change its position or orientation, it is still the same triangle.

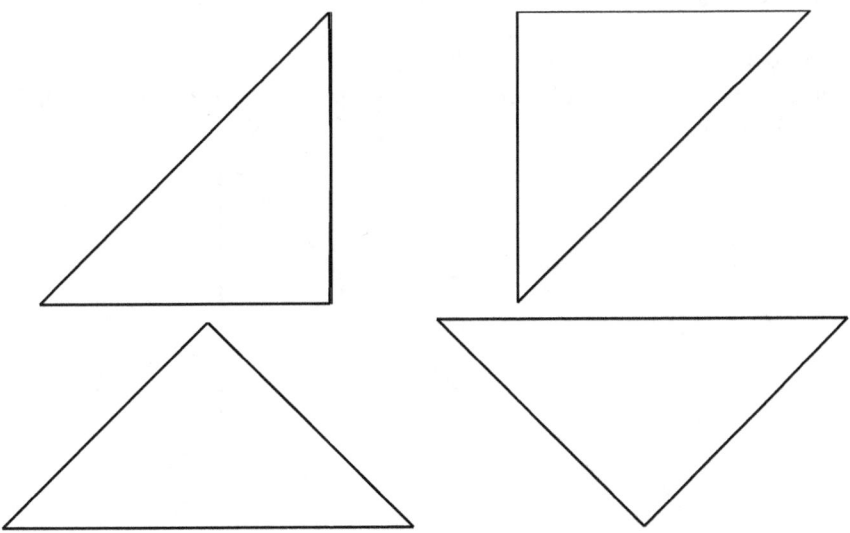

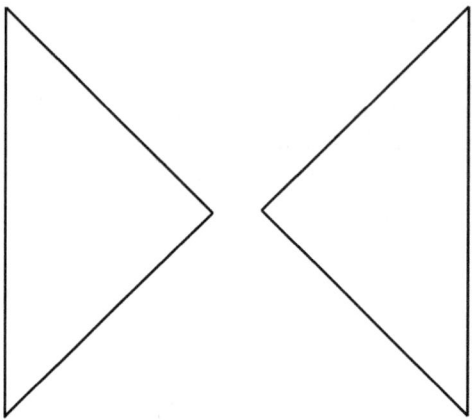

Figure 4

**Activity 3**
Children use two of the small triangles to cover the square in figure 1. They see that the amount of cardboard needed to make the square is twice as much as the cardboard needed to make the triangle, and that the area of the square is twice as big as the area of the triangle. They use the same two triangles to cover the square in figure 2.

**Activity 4**
Children take the big cardboard triangle, and cover each of the triangles in figure 5 with it.

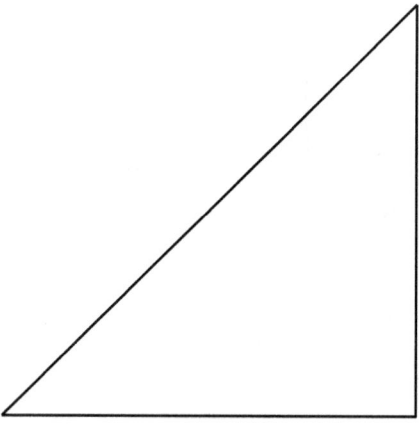

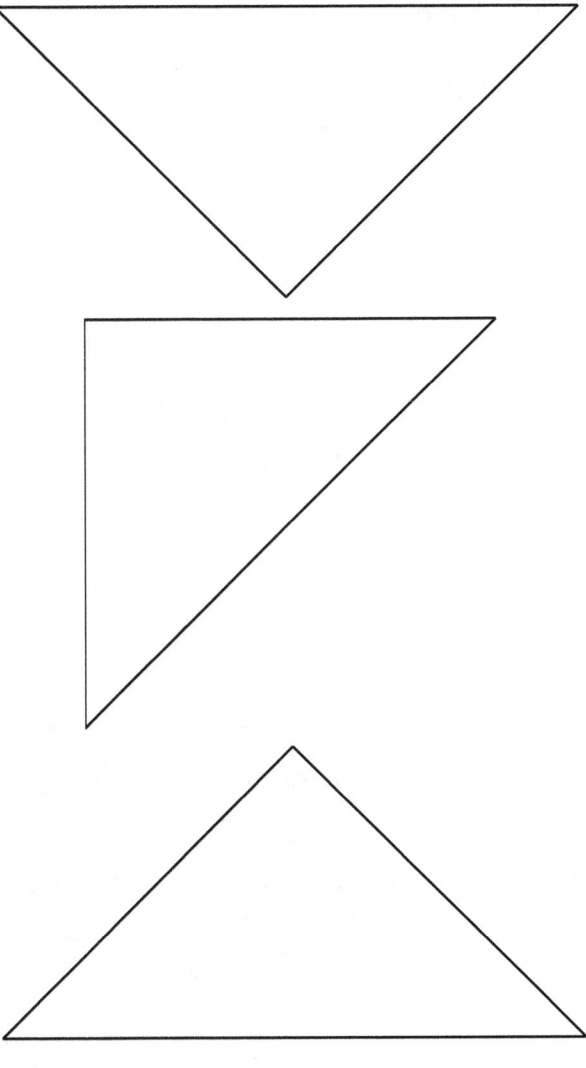

Figure 5

**Activity 5**
Children use two small triangles to cover each of the triangles in figure 5, and see that the area of the big triangle is twice as big as the area of the small triangle.

**Activity 6**
Children compare the area of the square with the area of the triangle, they cover the square (fig. 6) with two small triangles, and use the same two

triangles to cover the big triangle (fig. 6). They see that the area of the square is equal to the area of the big triangle. Although the square and the big triangle are different in shape, they have the same area.

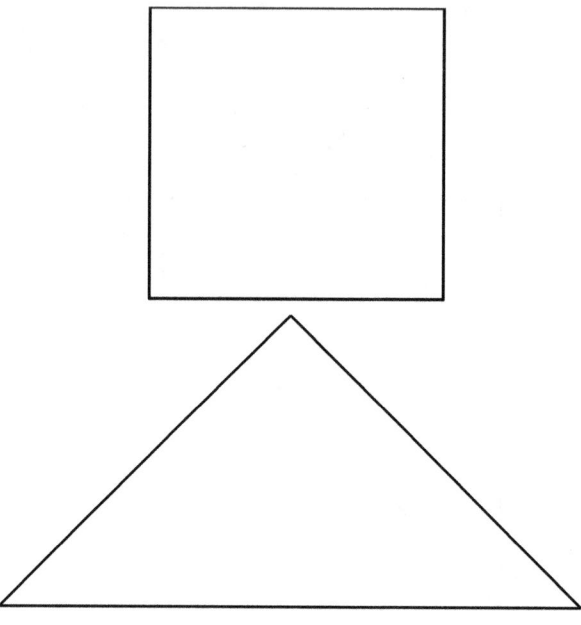

Figure 6

**Activity 7**
Children cover the shapes in figure 7 with two small triangles, to see that these shapes have the same area as the square.

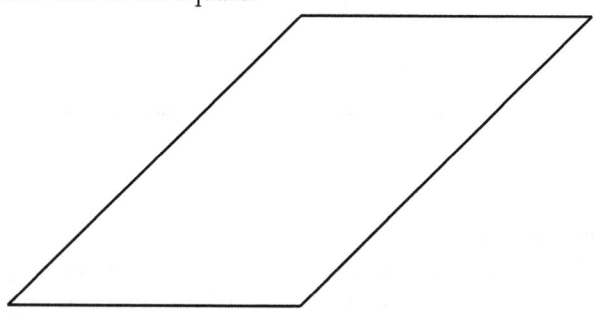

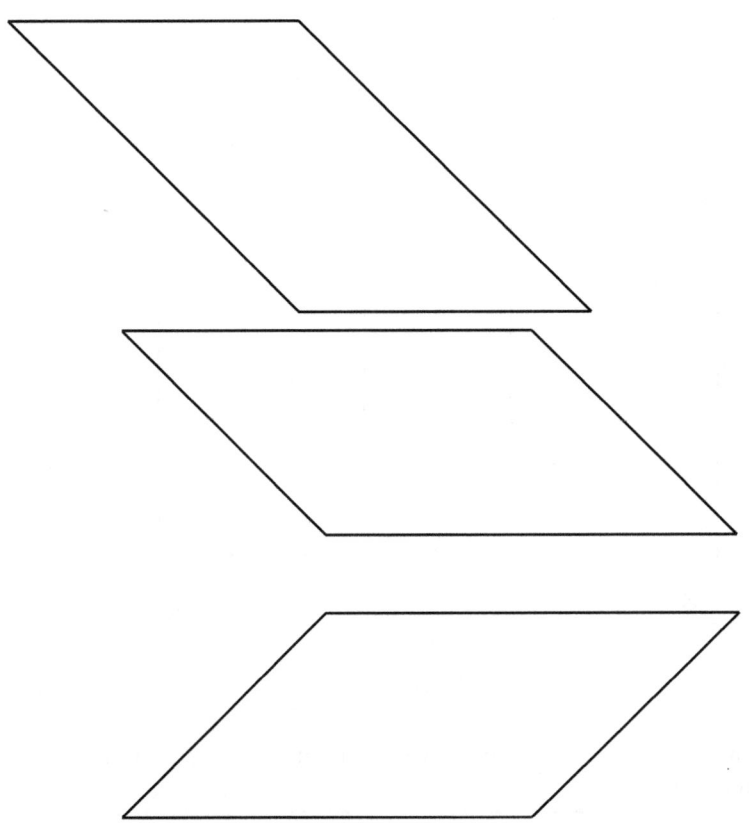

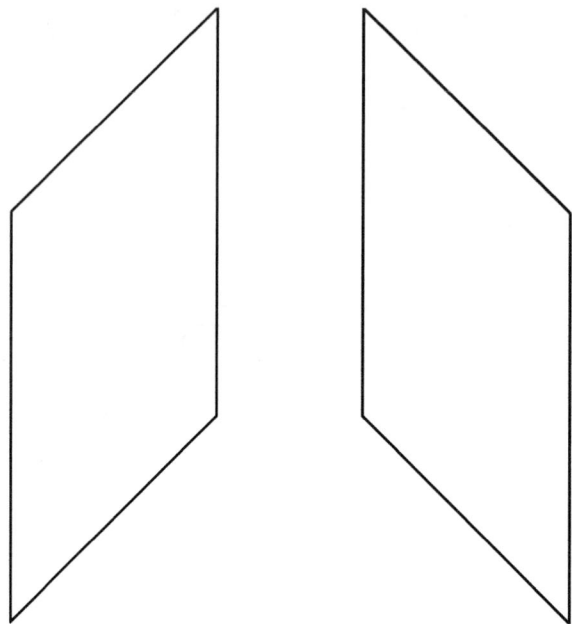

Figure 7

**Activity 8**

Children cover the square in figure 8 with four small triangles, and then the square in figure 9.

Figure 8

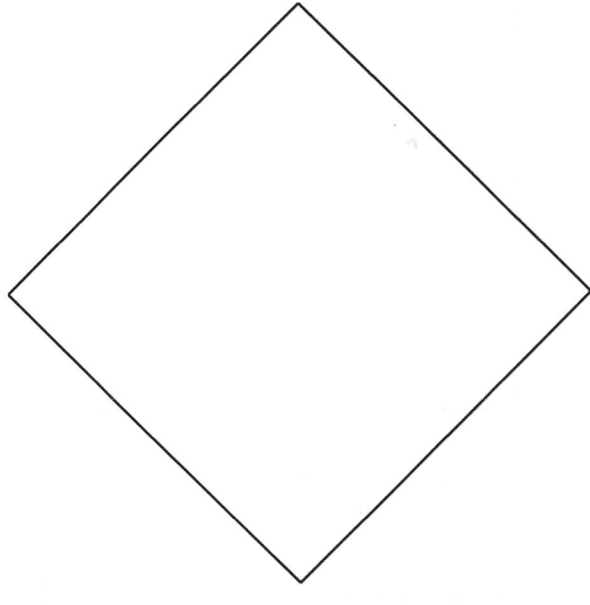

Figure 9

**Activity 9**
Children compare the area of the two small squares with the area of the big square. They cover each small square with two small triangles, and use the same four triangles to cover the big square. They see that the sum of the areas of the two small squares is equal to the area of the big square, and that the area of the big square is twice as big as the area of the small square.

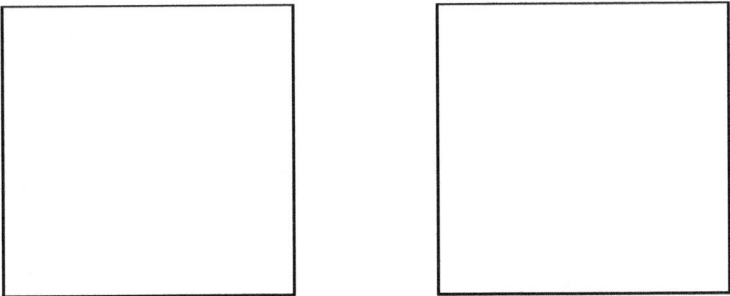

Figure 10

**Activity 10**

Children use the four small triangles to cover the two small squares in figure 11. With the same four triangles, they cover the big square. In this figure the sum of the areas of the two small squares is equal to the area of the big square. [The reader will recognize a special case of the Pythagorean Theorem.]

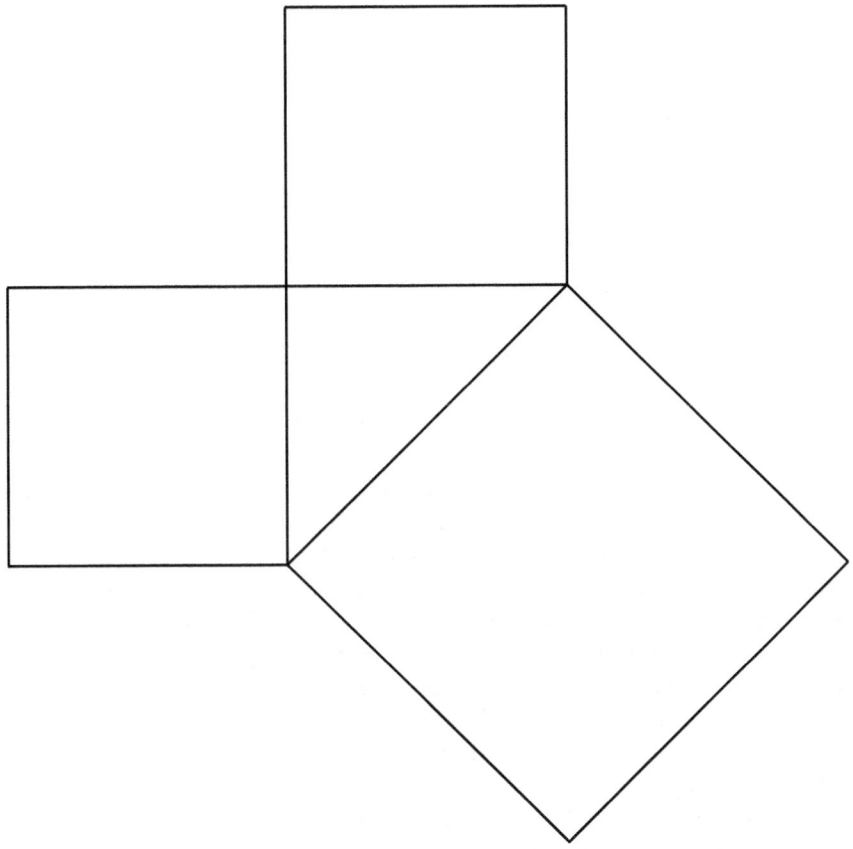

Figure 11

**Observations.**
In kindergarten, the children were familiar with the shape of a square. They were able to recognize and correctly show squares in their classrooms (calendar, boxes, frames, tiles, ceiling plafonds etc.) Only a few children pointed to some rectangles that were not squares. Many children first called figure 2 a diamond, although after the activity they were willing to call it a square, since it was clear that it was equal to the cardboard square and to the square in fig. 1.

All children did the activity of forming the square (in the two positions) with two triangles with relative ease. In this case matching equal angles or matching equal sides lead to the solution. Forming the big triangle took more time for some children, and some had to be helped, since they were using a strategy that was a dead end, matching corresponding angles from small and big triangles, so that there was no possibility to accommodate the second triangle (see fig 12).

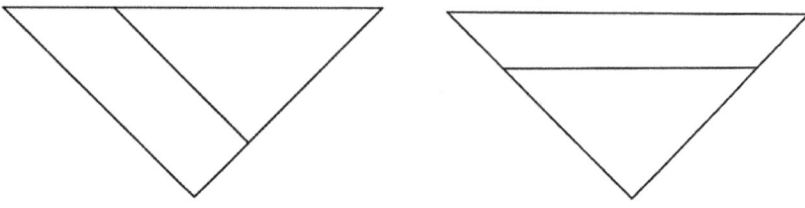

Figure 12

It was interesting to observe the different strategies used to form the same figure in different positions with the two triangles. Some children rotated the worksheet to get the new figure in a position similar to the figure they had solved first. Some children took the two small triangles together and translated and rotated both at the same time to fit into a new figure. Some children reassembled each figure from scratch, some of them always using the same method, some of them using trial and error each time.

When asked which figure would require more cardboard to be made (had more area), if the square or the big triangle, some children in second and third grade responded that the triangle, "because it is bigger." After pointing to them that both were formed by two small triangles, they agreed that the two needed the same amount of cardboard (had the same area).

Forming the big square with the four triangles was also challenging for most children, although some found the solution very quickly. Here again some of the students had to be helped since they insisted in matching equal (right) angles (see fig. 13), rather than trying to match equal sides to get the solution (see fig. 14).

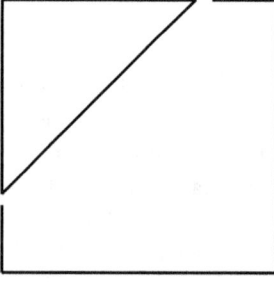

Figure 13

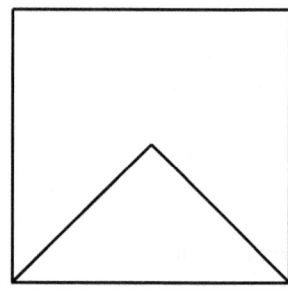

Figure 14

In a second grade, students were not given the frame in figure 8 to solve the problem "to form a big square with four triangles." Although it took more time in this class in general to solve the problem, the absence of the frame offered the opportunity for some children to find a different solution (see fig. 15)

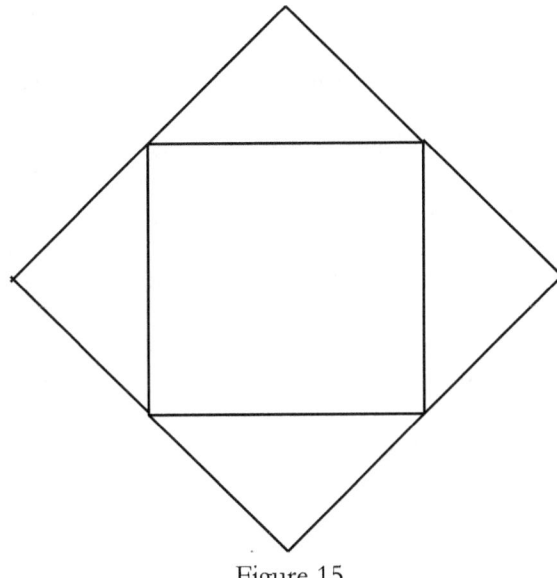

Figure 15

Students participated eagerly in the activities, some of them doing the activities ahead of the class on their own. Students liked the activities and were glad to keep the handouts.

Before the activities were conducted in the classrooms, a workshop on the Pythagorean theorem was conducted with the teachers, to give them a broader perspective of the main ideas of the activities. Classroom teachers were present during the class presentation, and most teachers participated, helping students and answering individual questions. Teachers reacted positively to the activities presented. Teachers were surprised by the difficulty that some of the children were having in forming the figures with the small triangles. After the activity, several teachers took out a tangram set to continue. Others asked for more materials and activities similar to these. Other teachers cut their own sets of figures, to continue the activities. In all, it was a positive experience in geometry for students and teachers in the early grades.

**References**

Challenge: The mathematically able student [Special issue]. <u>Arithmetic Teacher,</u> <u>28</u> (6). (February 1981).

Chancellor, Dinah. (1991). Higher-order thinking: a basic skill for everyone.

Arithmetic Teacher, 38, (6) p.48 - 50.

House, P. A. (Ed.) Providing opportunities for the mathematically gifted, K-12. National Council of Teachers of Mathematics, 1987.

National Council of Teachers of Mathematics. (1991a). Early childhood mathematics education. In 1991-92 Handbook. NCTM Goals, leaders, and positions. National Council of Teachers of Mathematics, p. 16.

National Council of Teachers of Mathematics. (1991b). Provisions for mathematically talented and gifted students. In 1991-92 Handbook. NCTM Goals, leaders, and positions. National Council of Teachers of Mathematics, p. 19-20.

National Council of Teachers of Mathematics. (1989). Curriculum and Evaluation Standards for School Mathematics. National Council of Teachers of Mathematics.

National Council of Teachers of Mathematics. (1991). Professional Standards for Teaching Mathematics. National Council of Teachers of Mathematics.

Pérez, Rosa Isela. (1991). Project Excel. San Diego City Schools.

# 9 CONNECTIONS IN PROPORTIONAL REASONING: LEVERS, ARITHMETIC MEANS, MIXTURES, BATTING AVERAGES, AND SPEEDS[9]

The lever is used to give an alternate physical representation of, and as a means to connect situations that involve weighted averages and inverse proportionality, such as arithmetic means, mixtures, batting averages, and speeds. Geometric representations of the situations are also provided as another way to make connections.

Consider the following situations:
1) Two children of unequal weight want to balance on a seesaw (see figure Where does the heavier child have to sit?

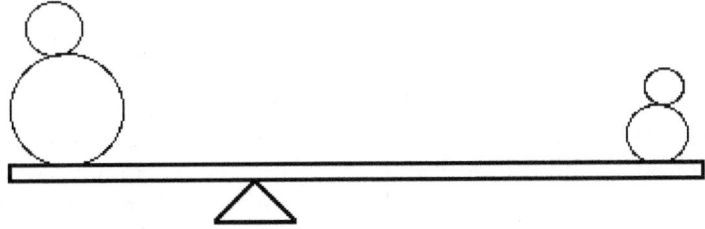

Figure 1. Two children of unequal weight want to balance.

2) A seller has two kinds of nuts; one costs 90 cents a pound, the other 60

---

[9] Flores, A. (1995). Connections in proportional reasoning: Levers, arithmetic means, mixtures, batting averages, and speeds. *School Science and Mathematics, 95*, 423-430. Used by permission of Wiley and School Science and Mathematics Association.

cents a pound. He wishes to make 50 pounds of a mixture that will cost 72 cents a pound. How many pounds of each kind should he use?
3) A baseball player has a batting average of .250 (15 hits in 60 at-bats) going into a game. That evening she gets 3 hits out of 4 at-bats. What is her new batting average?
4) A traveler drives the first half distance of his trip at 60 km/h, and the second half distance at 120 km/h. What is the average speed for the whole trip?

Making explicit what these apparently diverse topics have in common, and how they can throw light on each other, is one way to establish connections in mathematics and connections between mathematics and other subjects. The unifying theme of these problems is proportional reasoning, in particular, inverse proportionality and weighted averages, within multiplicative structures.

Proportional reasoning is one of the most important components of formal thought acquired in adolescence. Proportional reasoning has been described both as a cornerstone of algebra and higher-level mathematics and as a capstone of elementary arithmetic, number, and measurement concepts (Lesh, Post & Behr, 1988). Hoffer and Hoffer (1992) pointed out that failure to develop proportional reasoning precludes study in a variety of disciplines requiring quantitative thinking and understanding, including algebra, geometry, chemistry, physics, and some aspects of biology. They added, however, that proportional thinking is not a topic that is taught particularly well in schools, often fostering meaningless manipulation of symbols and formulas. The NCTM (1989) stated that proportional reasoning "is of such great importance that it merits whatever time and effort must be expended to assure its careful development. Students need to see many problem situations that can be modeled and then solved through proportional reasoning" (p. 82).

Freudenthal (1983) pointed out that the learning process must be conducted in such a way that sources of insight are not clouded during the process of algorithmization and automatization. Therefore, he recommended returning repeatedly to the sources of insight, both during and after the process of algorithmization and automatization. Many times, students deal separately and in a computational way with problems like those shown above. However, they can gain a deeper conceptual understanding of levers, averages, mixture problems, and speed, by connecting them. Formulation of these connections provides students with an additional approach with which to solve the problems, by relating them to more familiar situations or to more convenient representations. They can develop a physical meaning for quantities such as the arithmetic average, and an intuitive sense of the problem and the solution. This is a way to gain a deeper and more meaningful understanding of proportional reasoning

## The lever

The lever is a (weightless) beam that leans on a balancing point called the fulcrum. The distance between the fulcrum and the weight is the lever arm (see figure 2a). The product of this distance times the weight is the moment of force (also called the torque) with respect to the fulcrum. If the weight is to the right of the fulcrum, this product is the clockwise moment of force, if the weight is to the left of the fulcrum, the product is the counterclockwise moment of force (see figure 2b). For additional discussion on the lever see also Schiffer & Bowden, 1984.

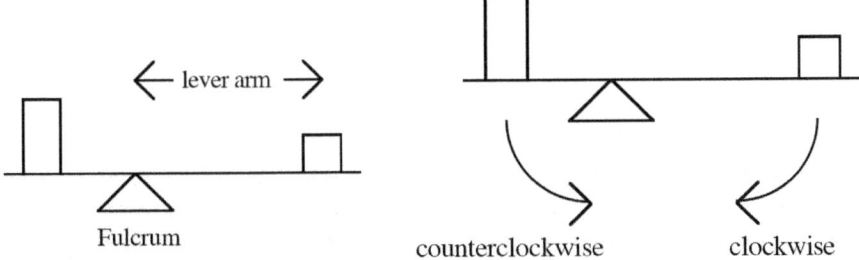

Figure 2a. Fulcrum and lever arm.    Figure 2b. Moments of force.

The lever is governed by the following basic principles:
1) Equal weights at equal distances are in equilibrium.

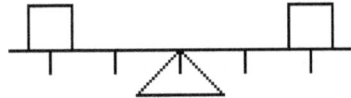

Figure 3a. Equal weights at equal distances.

2) The equilibrium of a (weightless) lever with a weight $w$ at each end will be undisturbed by replacing both weights by a single weight $2w$ at the midpoint. And conversely, $2w$ at the middle can be replaced by $w$ at each end without destroying the equilibrium.

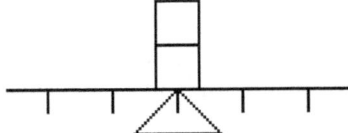

Figure 3b. Equilibrium is preserved.

From these two basic principles we can obtain the law of the lever (Pólya, 1977):

If a lever is in equilibrium, $w_1$ and $w_2$ are the weights, and $d_1$ and $d_2$ are their distances to the fulcrum, then $w_1 d_1 = w_2 d_2$ (the clockwise momentum is equal to the counter clockwise momentum).

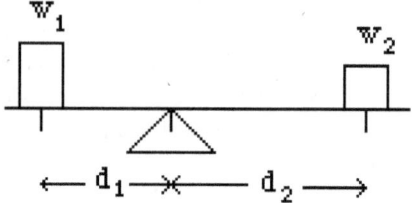

Figure 4. Weights and distances.

We can also rewrite this law as $\dfrac{w_1}{w_2} = \dfrac{d_2}{d_1}$
(the greater weight is closer to the equilibrium point than the smaller weight, so weights and corresponding distances are inversely proportional).

Students can relate this law to the familiar experience with a see-saw. Two children of unequal weights still can have fun if they put the see-saw to balance in a position closer to the heavier child (see Figure 1).

## The average and the lever

The arithmetic average of two numbers is simply the sum of the two numbers divided by two. Students can easily see that on the number line, the arithmetic average of two numbers corresponds to their midpoint, because $a + (b - a)/2 = (a + b)/2$ (see figure 5). Notice the similarity between the midpoint and the position of the fulcrum when balancing equal weights.

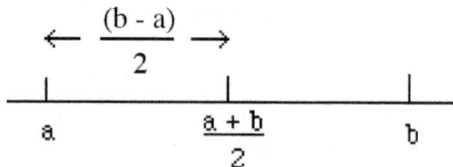

Figure 5. The arithmetic mean as the middle point.

In general, the average or mean $\bar{x}$ of a set of data $x_1, x_2, x_3, ..., x_n$ is the sum of those numbers divided by the total number or data, that is,

(1) $\bar{x} = \dfrac{x_1 + x_2 + \cdots + x_n}{n}$

In this case too, the mean of a set of numbers can be given a physical meaning. The number line is thought of as a weightless lever, and we put a unit weight on the numbers we want to average (see figure 6). The mean is the center of equilibrium of those numbers. The number line is thought as a weightless lever, and we put a unit weight on the numbers we want to average. Then, if we hold the lever at the mean, the system is in equilibrium (see Flores & White, 1989).

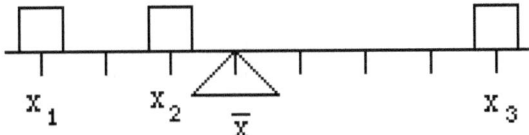

Figure 6. The arithmetic mean as center of equilibrium.

That this system is in equilibrium (clockwise momenta cancel the counter clockwise momenta) is equivalent to saying that

(2) $(\bar{x} - x_1) + (\bar{x} - x_2) + \cdots + (\bar{x} - x_n) = 0$.

The mean $\bar{x}$ of the values $x_1, x_2, x_3, \ldots, x_n$ is the only number that makes the sum (2) equal to zero. This characterization of the mean is equivalent to (1). For example, 6 is the average of 2, 4, 5, 8, 10 since

$(6 - 2) + (6 - 4) + (6 - 5) + (6 - 8) + (6 - 10) = 0$

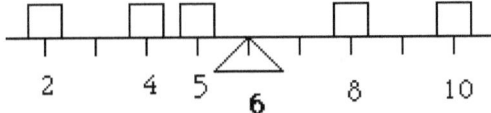

Figure 7. Arithmetic mean of numbers.

**Effect of extreme data on the mean.**

With the interpretation of the mean as the center of equilibrium, we can see why a single datum point that is very different from the others has so big an effect on the mean: the weight would have a very long lever arm. For example, if the data are 1, 2, 3, 4, 15, note that the mean $\bar{x} = 5$ is not very representative since it is greater than any of the data except for 15 (Fig. 7).

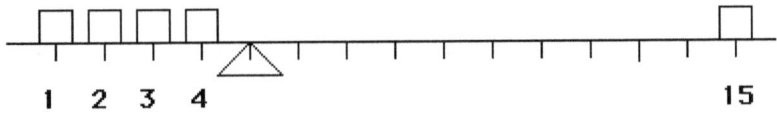

Figure 8. Effect of an extreme datum.

Notice too, that by dropping a single datum point (15), the average changes dramatically. The new average is 2.5 which is the midpoint of the remaining data. This sensibility of the mean to extreme data is the reason why other measures such as the median are necessary to describe a set of data.

## Weighted averages and a lever

For weighted averages the interpretation is the same. If different weights are put on various points of the lever, the weighted average of the values will be the point of equilibrium.

The number of times each datum appears is its "weight". The average will be closer to a datum that is repeated several times (has a greater weight). For example, the average illustrated in figure is 5 (since the weight at 4 is 5 times bigger that the weight at 10, the distance between 4 and the average has to be 1/5 the distance between the average and 10).

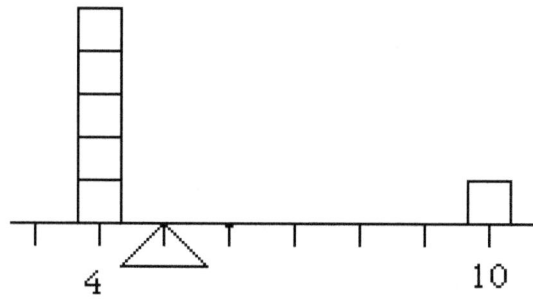

Figure 9. Weighted average.

That is, the average satisfies
$$w_1(\bar{x} - x_1) = w_2(x_2 - \bar{x})$$
or $w_1(\bar{x} - x_1) + w_2(\bar{x} - x_2) = 0$

This is equivalent to $\bar{x} = \dfrac{w_1 x_1 + w_2 x_2}{w_1 + w_2}$

To obtain the average we have to multiply each datum by its weight, and divide by the total number of unit weights.

The average of 4, 4, 4, ,4, 4, 10 is $\dfrac{5 \times 4 + 1 \times 10}{1 + 5} = 5$

Students can relate the weighted average to the see-saw. The position of equilibrium is precisely the weighted average of the positions of the two children.

For more data, the same idea works

The equilibrium point satisfies
$w_1(\bar{x} - x_1) + w_2(\bar{x} - x_2) + \cdots + w_n(\bar{x} - x_n) = 0$
This is equivalent to

$$\bar{x} = \frac{w_1 x_1 + w_2 x_2 + \cdots + w_n x_n}{w_1 + w_2 + \cdots + w_n}$$

For example, the mean of 2, 4, 4, 4, 5, 5, 8, 10, 10, 10 is
$$\frac{1\times 2 + 3\times 4 + 2\times 5 + 1\times 8 + 3\times 10}{1+3+2+1+3} = 6$$

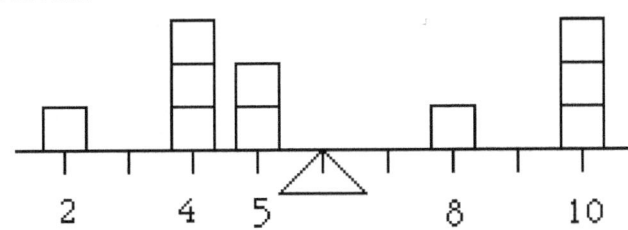

Figure 10. Weighted average of more data.

Realizing that this formula is not just a computational shortcut but a weighted average (see figure 10), can help students understand situations when averages have to be combined using the appropriate weights. Consider the following two problems:

a) A teacher wants to grade in the following way: Homework 20%, Quizzes 20%, Midterm 30%, Final 30%. How should she deal with the students' grades? What are the "weights" in this situation?

b) A student wants to compute his grade point average. Courses can be either 1, 2, or 3 credits, and grades range from 0 to 4. What "weight" should he give each grade?

## Mixture problems and a lever

On a lever a series of weights can be replaced by putting all the weights at the point corresponding to the point of equilibrium, without affecting the equilibrium of the lever. Similarly, we can replace a series of data by their (weighted) average. This idea will give us another method to solve mixture problems.

Example:
*A store owner mixes 12 kg of peanuts worth $4.00 / kg with 8 kg of cashews worth $9.00 / kg. At what price shall she sell the mixture?*

Let $p$ be the price. Since the weights are in the ratio $12/8 = 3/2$, $p$ must be the point between the values 4 and 9 that divides the segment in the ratio $3/2$ (and $p$ is closer to 4, since there is more weight there)

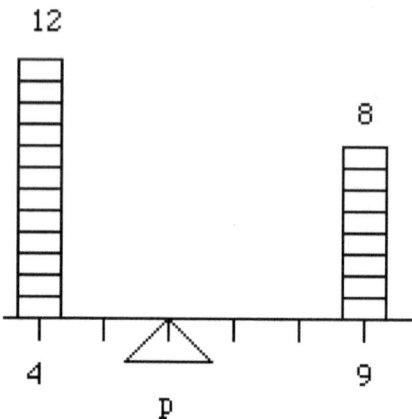

Figure 11. The price of a mix as a weighted average.

Other types of mixture problems can also be solved using the idea of a lever. The following problem is taken from Polya (1981).

*A dealer has two kinds of nuts; one costs 90 cents a pound, the other 60 cents a pound. He wishes to make 50 pounds of a mixture that will cost 72 cents a pound. How many pounds of each kind should he use?*

Let us represent the situation on a number line, thought as a lever. Here we have the inverse situation. We have all the weights at the equilibrium point 72, we want to put some weights at 60, and some at 90 without affecting the equilibrium.

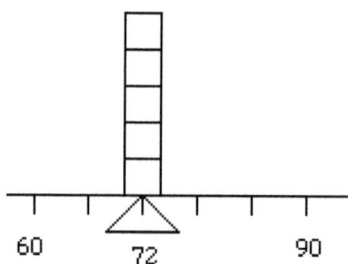

Figure 12. Price of mix is given.

Because the distance from 90 to 72 is larger than from 60 to 72 we have to put more nuts of the cheaper kind in the mixture, to balance the lever. To be more precise, the relation between $x$ and $y$ is given by $\frac{x}{y} = \frac{18}{12}$

So, the proportion of weights is 3 to 2. Complementing this information with the fact that the total weight is given by $x + y = 50$, we see that we need 30 pounds of the cheaper nuts and 20 pounds of the more expensive kind.

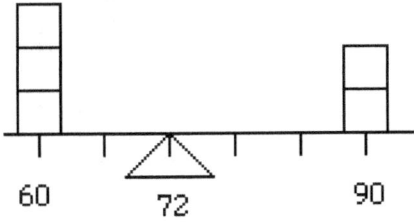

Figure 13. Solution to mixture problem.

## Batting Averages, Combination of Ratios

When computing a player's batting average, we do not just take the arithmetic mean of her previous batting average and the average of the day. If her batting average going into the game was .250 (15 out of 60) and she hits .750 (3 out of 4) during the game, the total number of hits is 18 and the total number of turns at bat is 64. The new batting average is 18/64 = .281.

The way the batting average is obtained is an example of how two ratios $a/b$ and $c/d$ are combined to form a new ratio (a + c) / (b + d). This new ratio will be between the other two:

If $\frac{a}{b} < \frac{c}{d}$ then $\frac{a}{b} < \frac{a+c}{b+d} < \frac{c}{d}$.

We can give a geometrical interpretation of this inequality (see Fig. 14).

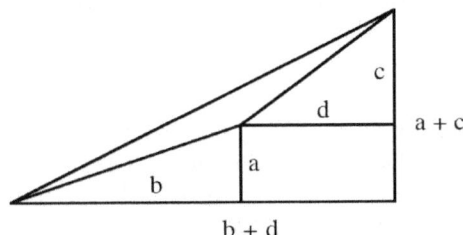

Figure 14. Combination of ratios as a slope.

If a and b are much larger than c and d respectively, then $\frac{a}{b}$ and $\frac{a+c}{b+d}$ will be very close together. The slope of the hypotenuse of the triangle with legs $a$ and $b$ is very close to the slope of the hypotenuse of the triangle with legs $a$ + $c$ and $b$ + $d$ (see Figure 15).

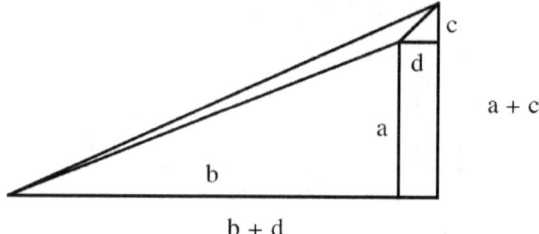

Figure 15. New slope is closer to "heavier" ratio.

This way to combine ratios is in reality a weighted average of the two ratios. In the case of the batting average example, what are the "weights" that correspond to .250 and .750? (see Figure 16).

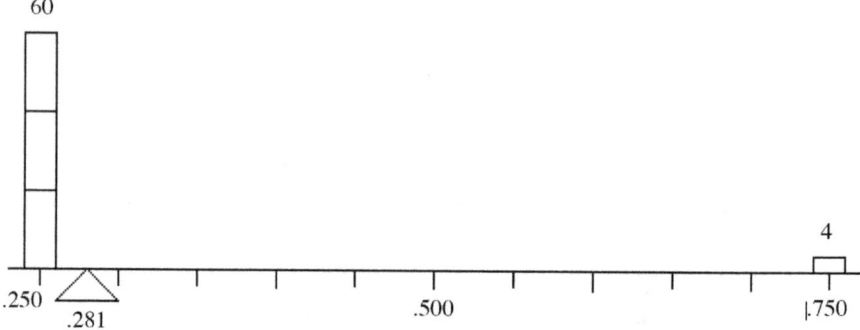

Figure 16. "Weights" of batting performances.

## Average of Speeds

Many drivers think that if they travel the first half distance of a trip at 60 km/h and the second half distance at 120 km/h, the average speed over the entire trip will be 90 km/h. To their surprise they arrive later than they think. That 90 km/h is not the average speed can be seen with one example. Suppose the driver travels 60 km at 60 km/h and then 60 km at 120 km/h. We get the average speed by dividing the total distance by the total time. For the first part of the trip the time is 1 hour, for the second part it is 0.5. The average speed is thus (60 + 60) km/ 1.5 h = 80 km/h. The amount of time required is not be the same in both cases, hence the slower speed is "weighted" more than the faster speed. Closely related to this problem is the one with an airplane flying a round-trip at constant speed in still wind, or flying a round trip with a tail wind one way and a head wind on the return trip. Another example is, of course, rowing upstream and back.

## Proportional reasoning in combining speeds

Sometimes students are unable to deal with the speed problem stated above because they are unable to compute the total time, as the distance is not given. We can think of the problem in terms of proportional reasoning, however, and use the same principle as used with the lever.

To obtain the average speed of an object traveling at two different speeds during equal times, the ordinary arithmetic average of the speeds is appropriate. If the speeds are unequal, we have to "weight" the speeds properly before averaging them. For example, if speed $s_1$ is twice speed $s_2$, for equal distances the traveler is going to spend twice as much time traveling at the slower speed. The average speed should be closer to the slower speed in a ratio of 1 to 2 (see figure 17). If the speed is 3 times greater, then the amount of time at the slower speed is going to be 3 times as big, and the average is going to be closer to the slower speed in a ratio of 1 to 3. In general, the amount of time spent at a given speed is going to be proportional to the inverse of the speed, and the average speed is going to be closer to the slower speed. When averaging speed over equal distances, the "weight" that we must attach to each speed is precisely the inverse of that speed.

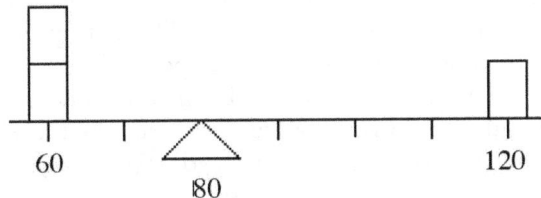

Figure 17. "Weights" of speeds.

## The harmonic mean

Another way to look at the problem is to compute the average speed over the whole interval. To do this, we compute the total distance and divide by the total time. If we are traveling at speeds $a$ and $b$, over equal distances $d$, the time required to travel the first distance is $d/a$ and the second is $d/b$. The average speed is then

$$\frac{2d}{\frac{d}{a}+\frac{d}{b}} = \frac{2}{\frac{1}{a}+\frac{1}{b}} = \frac{2ab}{a+b}.$$

The value $h = \dfrac{2ab}{a+b}$

is called the harmonic mean of a and b. Thus, to obtain the average speed of an object
traveling the same distance at two different speeds, the harmonic mean of the speeds has to be

used rather than the arithmetic mean. The harmonic mean $h$ of $a$ and $b$ is sometimes given as
$$\frac{1}{h} = \frac{1}{2}\left(\frac{1}{a} + \frac{1}{b}\right)$$
The harmonic mean $h$ of a and b can also be expressed as
$$\frac{a}{b} = \frac{a-h}{h-b}$$
We can make explicit the similarity of this formula with (3), where h is the "weighted" average of $a$ and $b$, with "weights" $1/a$ and $1/b$ respectively, by writing

$$\frac{\frac{1}{b}}{\frac{1}{a}} = \frac{a-h}{h-b}.$$

When averaging speeds over equal distances, we can see that in general the harmonic mean is smaller than the arithmetic mean, because the average speed will be closer to the slower speed than to the higher. When does equality of the two means hold? However, if the speeds are not very different, the arithmetic mean of the speeds will be a good approximation to the average speed. In car races on a closed race course, for example, a common procedure for determining average speed over different laps in a qualification match is to use the arithmetic mean of the speeds for each lap, instead of computing the total distance and dividing by the total time. What drivers will benefit more with this "incorrect" procedure? Will it be those whose speeds from lap to lap are close or those whose speeds vary widely?

## Other Means and Connections to Geometry

Another mean used in mathematics is the geometric mean, also called the middle proportional. The geometric mean of a and b is defined as $\sqrt{ab}$, or as the number $x$ that satisfies $a / x = x / b$. If we have a rectangle of side lengths $a$ and $b$, then the geometric mean is the side length of a square with the same area as the rectangle. Another representation of the geometric mean is given in figure 18. To show this, use the fact that the height on the hypotenuse of the right triangle divides the triangle into two triangles similar to it. You can also use figure 18 to show that $(a+b)/2 \geq \sqrt{ab}$, because the height of the triangle does not exceed the radius, which is $(a+b)/2$.

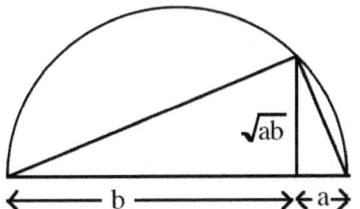

Figure 18. The geometric mean of *a* and *b*.

We can give another geometrical interpretation of the arithmetic mean, the geometrical mean, and the harmonic mean (Beckenbach & Bellman, 1975). In a trapezoid with bases *a* and *b*, the arithmetic mean corresponds to the mid-parallel of the trapezoid (see figure 19). The geometric mean $\sqrt{ab}$ corresponds to the length of the parallel that divides the trapezoid in two similar trapezoids. The harmonic mean corresponds to a parallel that goes through the intersection of its diagonals.

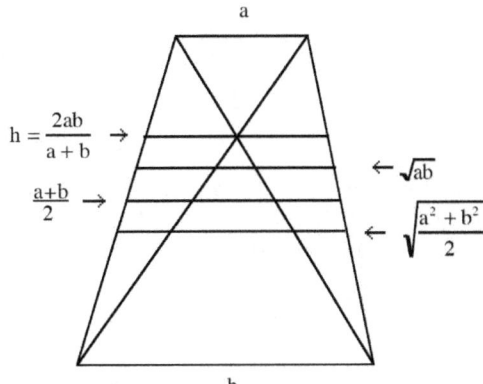

Figure 19. Harmonic, geometric, arithmetic, and quadratic means of *a* and *b*.

There is one more mean of two numbers a and b that is sometimes used in mathematics, the quadratic mean of *a* and *b* is defined as

$$\sqrt{\frac{a^2+b^2}{2}}$$

The quadratic mean corresponds to the length of the parallel that divides the trapezoid into equal areas. From the trapezoid (or with a little algebra for each inequality) we see that we have this relation between the harmonic, geometric, arithmetic, and quadratic means:

$$\frac{2ab}{a+b} \le \sqrt{ab} \le \frac{a+b}{2} \le \sqrt{\frac{a^2+b^2}{2}}.$$

A good exercise for students who have dealt with similar triangles is to prove that the different means of $a$ and $b$ do correspond to the mentioned parallels in the trapezoid — another way to practice proportional thinking.

## Conclusion

In this article we made explicit the underlying mathematical structure of the four different problem contexts introduced at the beginning (weighted averages and inverse proportions). We provided alternate representations of these concepts (physical, algebraic, and geometric), as they are related to one another. By helping our students make these connections, they can develop a conceptual understanding, that goes beyond the mere computational ability required to solve each problem separately. However, weighted averages and inverse proportions are only one aspect of proportional reasoning, and it is beyond the scope of this article to explore all aspects of this field. Mochon (1993) provides a description of a broader range of cases, discussing how ratios, rate, and fractions can be added in a meaningful way. As he points out, the problem context will provide the rationale for the approach to operate on the numbers.

## References

Beckenbach, E. & Bellman, R. (1975). An introduction to inequalities. Washington, D.C.: Mathematical Association of America.

Flores, A., White, A. L. (1989). Exploration of the mean as a balance point. School Science and Mathematics, 89, 251-258.

Freudenthal, H. (1983). Didactical phenomenology of mathematical structures. Dordrecht: D. Reidel.

Hoffer, A. R. & Hoffer, S. A. K. (1992). Ratios and proportional reasoning. In T. R. Post (Ed.),

Teaching mathematics in grades K-8: Research based methods (p.303-330). Boston: Allyn and Bacon.

Lesh, R., Post, T. & Behr, M. (1988). Proportional reasoning. In J. Hiebert and M. Behr (Eds.), Number concepts and operations in the middle grades (p. 93-118). Reston: NCTM.

Mochon, S. (1993). When can you meaningfully add rates, ratios and fractions? For the Learning of Mathematics, 13 (3), 16-21.

National Council of Teachers of Mathematics. (1989). Curriculum and evaluation standards for school mathematics. National Council of Teachers of Mathematics.

Pólya, G. (1977). Mathematical methods in science. Mathematical

Association of America.

Pólya, G. (1981). <u>Mathematical discovery: On understanding, learning and teaching problem solving</u>. Wiley.

Post, T. R., Behr, M. J., & Lesh, R. (1988). Proportionality and the development of prealgebra understandings. In A. F. Coxford and A. P. Shulte (Eds.), The ideas of algebra, K - 12. 1988 Yearbook (p.78-90). Reston, Va.: NCTM.

Schiffer, M. M. (1984). <u>The role of mathematics in science</u>. Mathematical Association of America.

# 10 ORCHESTRATING, PROMOTING, AND ENHANCING MATHEMATICAL DISCOURSE IN THE MIDDLE SCHOOL: A CASE STUDY[10]

"The discourse of a classroom - the ways of representing, thinking, talking, agreeing and disagreeing - is central to what students learn about mathematics"
*Professional standards for teaching mathematics*, p. 34.

## Mathematical discourse in the classroom

To have a real understanding of mathematics, students should be able to form internal, mental representations of mathematical concepts, and to build connections among them, and also between those representations and external representations such as concrete materials that embody the abstract mathematical concepts, and between the representations and the symbolic notation used for those concepts (Hiebert & Carpenter, 1992). Communication of these representations and their connections is essential to see whether the student has attained understanding. However, there is also another reason that makes communication very important in mathematics. As Cazden (1986) highlights, talking about the representation provides the opportunity to reflect on those representations. And as Pirie (1988) points out, students learn mathematical concepts not from their experiences with concrete materials, but by their reflections on those experiences. The importance of the relation of language and thought is of course not a recent discovery. Piaget (1932) stated that children's participation in social conflict such as

---

[10] Flores, A., Sowder, J. T., Philipp, R., & Schappelle, B. (1995). Orchestrating, promoting, and enhancing mathematical discourse in the middle school: A case study. In J. T. Sowder & B. P. Schappelle (Eds.) *Providing a foundation for teaching middle school mathematics* (pp. 275-299). Albany, NY: State University of New York Press. Used by permission.

arguments enhances the development of perspective-taking skills, stimulating cognitive growth. Vygotsky (1934/1962) stressed the role of social language in the development of thought. Bishop (1985) stressed the importance of the social construction of meaning for mathematics education.

Enhancement of the quality of mathematics teaching and learning in the classroom requires consideration of several aspects. NCTM (1991) discusses the following standards for teaching mathematics: Worthwhile mathematical tasks, mathematical discourse, learning environment, and analysis of teaching and learning. All of these are important and essential, and teachers need to experience growth in all of them. However, mathematical discourse is a key element. While stressing mathematical discourse, a teacher can in a natural way consider the other aspects of teaching mathematics. That is, the activities chosen by the teacher can be worthwhile mathematical tasks, they can be conducted in a learning environment that fosters higher-order mathematics thinking, and teachers can analyze their teaching practices and the learning process.

The ways of representing, thinking, talking, agreeing, and disagreeing about mathematics constitute the classroom discourse (NCTM, 1991). *Discourse* is used to highlight the ways in which knowledge is constructed and exchanged in classrooms. Who talks? About what? In what ways? What do people write down and why? What questions are important? Whose ideas and ways of knowing are accepted and whose are not? What makes an answer right or an idea true? What kinds of evidence are encouraged or accepted? (Ball, 1991). There are three components in mathematical discourse: 1) The teachers' role in discourse; 2) The students' role in discourse; and 3) Tools to enhance mathematical discourse. As NCTM stresses, all three are very important. The possibility of making available for reflection the processes by which students relate new knowledge to old through discourse, depends on the social relationships, the communication system, which the teacher sets up (Cazden, 1986). Mathematics learning is both a process of individual construction of knowledge and meaning and a process of acculturation into the mathematical meanings and practices of wider society (Eisenhart, 1988). As Cobb, Yackel and Wood (1992) say, it is the teacher's role to facilitate mathematical discussions between students while at the same time acting as a participant who can legitimize certain aspects of their mathematical activity and sanction others. "The teacher's role is to facilitate the development of this type of mathematical discourse by helping students in their attempts to express their mathematical thinking while also encouraging them to conceptualize situations in alternative ways. ...The teacher facilitates the student's mathematical development by subtly highlighting selected aspects of their mathematical contributions" (Yackel, Cobb, Wood, & Merkel, 1990, p. 35).

One of the reasons that the students' role is so important in discourse is that students have to take an active part not only in doing but on reflecting on what they do to construct their understanding. Discussion both focuses these reflections and makes it possible to share them. Discussion, as it is used here, "is purposeful talk on a mathematical subject in which there are genuine pupil contributions and interactions." (Pirie, 1988, p. 2). As Yackel, Cobb, Wood and Merkel point out, collaboration involves much more than combining solution procedures to develop a joint solution. It involves developing explanations that are meaningful to someone else and trying to interpret and make sense of another's ideas and solution attempts as they evolve. In this process, students try to verbalize and interpret ideas that are only partially formed. During subsequent whole-class discussion, students are expected to give coherent explanations of their problems, interpretations, and solutions and to respond to questions and challenges posed by other students. They are also expected to listen to, and try to make sense of, explanations given by other students, to pose appropriate questions, and to ask for clarifications. Yackel, Cobb, Wood and Merkel continue saying that when students engage in this type of discourse, not only is the amount of time they spend participating in problem-solving activities increased, but the nature of their problem-solving activity is itself extended to encompass learning opportunities that rarely arise in traditional instructional settings. As students participate in the discourse they learn to develop mathematical explanations and justifications. Tools such as concrete materials, calculators and computers provide the necessary common experiences to communicate. Ball (1991) states that without explicit attention to the patterns of discourse in the classroom, the long-established norms of school are likely to dominate - competitiveness, an emphasis on right answers, the assumption that teachers have the answers, rejection of nonstandard ways of working or thinking, patterns reflective of gender and class biases. In too many mathematics classrooms, answers have traditionally been right because the teacher says so. Too often, students are unable to define mathematics vocabulary using their own language (Miller, 1993).

During the interaction between the students and the teacher, not only the official mathematical knowledge is constituted, but also the mathematical competence on the part of the student is established (Jungwirth, 1991). Jungwirth has documented that teachers have gender-specific practices. Modifications of the typical patterns result from gender-specific student practices. Modifications also result from the actions of the teachers in correspondence to the students' actions.

Mathematical discourse helps to make mathematics relevant, by helping to establish the necessary connections between the internal and external representations and students' experiences. And as Lampert (1989) stresses, this connection makes it possible to shift the locus of authority from the teacher and the textbook towards a classroom community (teacher and students) that has the power to use mathematical tools to decide whether an answer or a procedure is reasonable.

By stressing classroom discourse the teacher can addresses the special needs of all students. Discourse can help remedy difficulties not by working at the symptoms of the problem, but by working at the root of the problem. To do mathematics, it is necessary to manipulate certain mental objects, namely mathematical concepts, using symbols as combined labels and concepts. But as Skemp (1982) points out, for many children and adults these mental objects are not there, and they learn to manipulate substitute objects "empty symbols, ... labels without content". The problems that so many have with mathematical symbols arise largely from the absence of the concepts and relations that give symbols their meaning. The remedy lies in the building up of the conceptual structures. Skemp offers several suggestions

    i) Give children many physical embodiments as possible of the mathematical concepts which we want to help them construct

    ii) Sequence the presentation of new material in such a way that it can always be assimilated to a conceptual structure, and not just memorized in terms of symbolic manipulations

    iii) Stay longer with spoken language. The connections between thought and spoken words are initially much stronger than those between thought and written words or symbols.

    iv) Use informal, transitional notation as bridges to the formal, highly condensed notations of traditional mathematics.

Discourse makes explicit the fact that each child learns differently, that individual students interpret instructional situations in profoundly different ways (Yackel, Cobb, Wood, & Merkel, 1990). Because of the variety of ways that students can use to represent, think, talk, and write about mathematics, emphasis on discourse is a strategy that is appropriate for all students. Discourse can help understand cultural differences that minority students bring to the classroom. Discourse provides opportunity for bilingual students to develop their communications skills in mathematics. The mathematical discourse that goes on while students work in small groups also promotes cooperation among student and encourages them to take an active participation. Some evidence has been found that explaining enhances learning by the explainer (Lindow, Wilkinson, Peterson, 1985). Wood and Yackel point out that peer group interaction gives rise to learning opportunities by encouraging both an exchange of viewpoints and verbal elaboration. Children who are committed to collaborate, try to make sense of each other's interpretations of the situation at hand and engage in mutually supportive activity. Also, when children attempt to make sense of each other's problem-solving attempts, they must extend their own conceptual framework to try to construct a consensual domain with their partner. According to Wood and Yackel this may occur when partners arrive at the same answer via different solution methods, as well as when one child tries to explain a perceived error in another's thinking. Additionally, language can both help children reflect on their own understanding when they give explanations and help them re-conceptualize their own cognitive constructions as they attempt to make sense of their partner's explanations.

## Introduction

Jo is a gentle and quiet person. She teaches all subjects in a fifth-grade classroom with 33 students in an urban school with students coming from a variety of ethnic and cultural backgrounds and wide range of incomes. Although Jo is an experienced teacher, this is her first year at that school. She is one of the four extraordinary teachers that were described earlier (Philipp, Flores, Sowder & Schappelle, 1992). During the seminar sessions that were part of the study, she participated in her own style, being on occasions almost self-effacing. Jo is sometimes also a self-effacing teacher, in the same way that some outstanding orchestra directors are. This metaphor, the teacher as an orchestrator, used in the *Professional Standards for Teaching Mathematics* (NCTM, 1991), will serve as the leitmotif for this case study. In this paper, we will focus on one of the most salient, extraordinary features in Jo's teaching, the way she orchestrates, enhances and promotes the mathematical discourse in her classroom. Jo's main objective, is to teach for understanding. In her own words: "My primary focus is understanding. I want those kids to know not just how to do it, but why it works." Of course, many teachers' main purpose, is teaching for understanding. However, as Hiebert and Carpenter (1992) point out, although widespread, this objective has been difficult to attain. We will describe the role of mathematical discourse in Jo's classroom to attain this objective. We will also use the Standards frame for describing teacher's and students' role in discourse, and for tools for enhancing discourse described in that book. From the mathematical point of view, the study in which Jo took part had its main interest in the concepts of rational numbers multiplicative structures in the middle grades. The data were collected through interviews, participation in seminars, written answers to questions, and extensive observations of her practice in the classroom, several visits during the year, and one continuous week. Class observations were made most of the time by two observers, and were audio-taped. All students quotes come from classroom observations, and Jo's quotes are from the interviews or from the classroom (the context shall make it clear).

## Teacher's role in discourse

Jo stated in the interview that she perceives the teacher's role in the classroom "in large part as a facilitator." She tries "to lead the kids and hopefully get them to figure things out on their own."

Posing questions and tasks that elicit, engage, and challenge each student's thinking

Jo frequently asks questions that do not require a single number for an answer, but that elicit students' thinking and requires them to express in their own words relations between mathematical concepts. For example, in the discussion about factors she asked these questions: "What can you say about factors?" "What do you notice about the factors of a number?" "What do you notice about the multiples of a number"

The tasks that Jo chooses for her students are challenging and engage them in mathematical thinking. For example, in the factor game (see Fitzgerald et al., 1986), students analyzed which numbers were a good option to start with (depending on the number of factors). Students were involved and participated actively in the game, both with the class as a whole and when playing in pairs.

Jo has a math corner where she has different kinds of problem solving activities. She has about three different activities which she changes fairly regularly. She encourages the kids when they have time to go to those and use them as a challenge. She said "I do not give the kids that are more able or faster just more to do, I try to make things that are interesting to them."

Listening to students' ideas

Jo listens carefully to what students say, both to correct and incorrect ideas. She does not just ignore incorrect answers; she addresses misconceptions expressed by her students using many times what other students said. For example, answering the question what were factors, a student said: "multiples of other numbers". Jo did not say anything immediately, but when another student said "15 is a factor [of 30], you can multiply to 30, $15 \times 2$" Jo asked the first student: "Is 15 a multiple of 30?"

The fact that Jo listens to her students is also reflected on how she plans and changes her plans. Jo stated in the first interview that she always has a clear objective for each class "I have a basic outline for what I am doing for several days, maybe a week, but I may erase everything. I always make my plans in pencil, because I may have to, based on what the reaction to a lesson is on one day or even part of a day, change everything around."

Asking students to clarify and justify their ideas orally and in writing

Jo asks students to keep a mathematical diary. She asked them to write about factors and then share their writings with the classroom orally.

During the interview Jo said "If I ask them good questions, it is the biggest way to tell what they really understand and what they do not. Sometimes that can be a real eye opener." She added that you can find a lot easier what the kid understands or where the misunderstanding is "by having them explain something than just looking at an answer on a test."

<u>Deciding when and how to attach mathematical notation and language to students' ideas</u>
Mathematical notation is introduced in a natural way once the concepts and ideas have been discussed. The meaning of the notation is made explicit, and the relations between closely related or alternate notations is discussed. For example, during a lesson on division the different ways to express the result, as a quotient with or without remainder, as a decimal, or as a mixed fraction were introduced and related to each other.

<u>Deciding when to provide information, when to clarify an issue, when to model, when to lead, and when to let a student struggle with a difficulty</u>
In Jo's class, students frequently are engaged in a single rather difficult, multi-step problem for a good part of the hour. Jo lets students struggle with the difficulties, individually, in pairs or in small groups, according to the type of problem. Students do not find struggling with problems as something out of the usual. They share their solutions and processes, think in loud voice, express their doubts. Jo intervenes during these problems, by guiding the students. For example, in a particular problem that required the multiplication of numbers with several decimals, students were spending time to multiply the numbers by hand, so she suggested to use the calculator so that they would not get bogged with the computations and could focus on the major steps of the problem.

<u>Monitoring students' participation in discussions and deciding when and how to encourage each student to participate.</u>
Jo encourages all students to participate. Students are praised for giving their own explanations of the concepts. Wrong answers are not given a negative response, but the misconception are clarified. Jo encourages students to extend their explanations, she gives them more time.

**Student's role in discourse**
Working with manipulative materials, being active with their hands is not enough for mathematics concepts to form and develop in students. As Pirie (1988) points out, students' concepts are formed, not by their experiences but by their interpretation of their experiences. Students usually are not used to take an active role in the mathematical discourse. It is hard, even for teachers like Jo to change some patterns of passive behavior that have developed over the years. In the first interview, Jo stated: "I want those kids to know not just how to do it, but why it works. And sometimes they get tired of that and they want the shortcut."

<u>Listen to, respond to, and question the teacher</u>
During the interview, Jo said that he kids seemed to have a pretty good

understanding of lining the values in places, when discussing addition of decimal numbers. "They told me, rather than the other way around."

### Listen to, respond to, and question one another
Jo tries that the mathematical discussion is not only between teacher and student, but she tries to promote the interaction of students. Jo has students very often, as she said in the interview, "do groups-of-four type of things, and come to a consensus about problems ... to discuss within themselves what they are doing and why."

Part of what Jo strives to do is to develop in students respect with what other students say, and respect the time they need to think. When one student became impatient with another's slow response, Jo told the impatient student to give the other student time to think.

However, it is not an easy task, students seem more eager to share their own explanations than to listen to others. Although most of the time students raised their hand to answer questions or explain solutions, sometimes students shouted their answers without waiting to be asked, sometimes they interrupted other students.

### Use a variety of tools to reason, make connections, solve problems, and communicate
Sometimes, students on their own initiative take out calculators when solving problems. They are confident using diagrams to share their solutions and procedures to their peers, use the overhead projector to communicate with them.

### Initiate problems and questions
Students frequently open discussions with their questions. An example will be seen towards the end of the excerpt of one of Jo's lessons, where a student initiated a rather lengthy discussion when he asked to clarify why when you multiply by 0.5 you get a number smaller than that you started with.

### Make conjectures and present solutions
Jo's students make conjectures. When discussing multiples and factors one student asked: "if it is a multiple, does it always have to be bigger?"

### Explore examples and counterexamples to investigate a conjecture
Students are asked examples when they make statements. They also give examples on their own initiative, in natural way. A student after being convinced by the discussion that $10 \times .5$ should be 5, offered his own example: $20 \times .5 = 10$.

Try to convince themselves and one another of the validity of particular representations, solutions, conjectures, and answers
Students do a problem of the week, and they get 10 possible points, but only two points are for a correct answer, the rest is, as she said in the interview "for their reasoning, putting it down, diagraming it, writing it, explaining it, however they can best convey it to me." As will be seen in the excerpt, students present (and defend) their methods and solutions to the group. They show poise and confidence when explaining how and why their method works.

Rely on mathematical evidence and argument to determine validity
Jo works very hard to develop in their students the idea that the teacher is not the only nor the main source of mathematical authority. Jo lets students share their answers and asks them to decide whether the answer or the process is appropriate. For example, during the discussion on factors, students asked questions and instead of answering herself, the questions were directed back to the students: "What do you think of that?", "Do you agree with that?", "Is that right?", "How can you prove your point?" "I am not going to say 'they are right [or] they are wrong'".

Relying solely on the authority of the textbook is also not enough. When a student justified his answer by saying that the textbook had an example like that, Jo mentioned that textbooks sometimes have mistakes. Students frequently use the calculator to verify an answer. However, some students are not satisfied by this only, they want to know why the result given by the calculator is reasonable. The student with the question of why multiplying by 0.5 makes the result smaller had obtained his answer with the calculator.

Jo uses cooperative groups frequently. She has posted three Cooperative Group Guidelines on the wall. The third is *3) No one may ask a question to the teacher unless everyone else in the group has the same question.*

## Classroom discourse. Some examples

The following problem was assigned to students to work in small groups.
*One farmer earned $12,600 from crop sales in 1984. Each year, she earns 0.04 times more than she did the previous year. How much will she earn in 1987.*
Next day, Jo asks members of the teams to "come up and explain to us how they did the problem"
One student explains the following strategy, multiply 0.04 times 3 to get 0.12, then multiply 0.12 by 12600 (=1512) and add this result to 12600.
S1: We came up with fourteen thousand eleven hundred dollars.
Jo: Okay. You didn't just come up with that though.
S2: Is that the right answer?
Jo: Well. we're gonna discuss that too.
S2: We didn't get that.

# TO CONNECT IS TO UNDERSTAND MATHEMATICS 4

Jo: Okay. Some of you may have gotten something different, and we're gonna talk about that.

Another student explains a different strategy.

Matt: First we multiply. We took 12600 and multiplied by zero four. And got 504. Then we added 504 to 12600, to get 13,104...

Matt carries the process on the overhead projector the (he had done the multiplications with the calculator).

```
   13104
    0.04
  524.16
   13104
13628.16
    0.04
 545.1264
13628.16
14173.286
```

Rounded to $14173.29

Jo: If you do it on a calculator, that is fine, but you need to tell people what you're doing.

Matt: Okay! I'm going to do it first.

Jo: [to the class] And if you have your calculators, press them. You should all have those calculators... [to Matt] you just need to let us know what you are doing.

Matt: Then we added 504 to 12,600. And then we got 13,104.

Jo: Okay. Now why did you add those together?

Matt: To get the total for '85. That is how much more she made.

Jo. Okay.

Matt: Then we multiplied 13,104 times 0.04 ... and we got 526 point 16. That's cents.

Jo: Okay.

Matt: That is the multiplication. Then we added 524.16 to our original which was 13 thousand.

S: Added?

Matt: Yeah. We added and got out 13,628.16

S: Is that the right answer?

S2: We are not done yet. One more step.

Matt: It is like ... a pattern ... And we came out with that for 1986.

Jo: Okay.

Matt: So for 1986 we got 15, ...

S: Is that right?

Jo: I want you to think about it and we're gonna discuss it. I'm not gonna

say, "They're right; they're wrong." I want you guys to talk about it.
Matt finishes explaining the problem.
Jo: Okay. Wait a minute now. Do we need to round off to the nearest penny?
Matt: That's what we're doing. We're just showing you.
Jo: I'm sorry. I should just be totally quiet.
[....]
Matt: So, that is 1 4 1 7 3 point 29 cents.
Ss: Yeah (applause)
S: Is that right?
Jo: Did anyone else agree with them?
S: We did.
Jo: You did? ... Helen, what do you think?
Helen: Because they multiplied it correctly and each step makes it ...
Jo: Chukie?
Chukie: I agree, 'cause me and Chris did it first and we got the same answer.
[...]
Jo: Why did you do those same steps that they did?
Chukie: Because we had seen we couldn't do it the same way that Una and Andrea did. They were trying to do it fast. But we did it that way first, but it didn't really work out... We got the answer.
Jo: Anybody else want to comment on this? Elsie.
Elise: Well, I think they did it correctly because we did it another way, and got the same answer.
Jo: Okay. How did you do it Elise? [...]
Elise writes on overhead projector

$$\begin{array}{rrr} 126 & 126 & 12600 \\ 100\,|\,\overline{12600} & \times\ 4 & +\ 504 \\ & 504 & 13104 \end{array}$$

Elise: First I knew that the 0.04 is also like 4%, so to find out how much each percent is worth in the 12,600, I just divided that by 100 and I got 126. Okay. So, then I just multiplied that 126 by four and I got 504. Okay. And then you can just add that to 12600. So, then you're gonna do that all over again. You have to divide 13,104 by 100; then that is - do you have your calculator?
Elise finishes explaining how they got the same answer.
Elise: It was $14,173.29
S: That's what they got?
S: Is that the answer?
Jo: How many of you did come out with that same answer, whether you did it one way or the other way? How many of you came out with a different answer? [...]
Jo: How many of you figured that they made this many times more money by doing this [0.04 × 3 = 012] and then calculated it out? How many of you did it in three separate steps? [...] How many of you did it a totally different

way?

Teacher discusses another example with the class, with easier numbers, starting with 1000 000 and 0.05 more times each year. The group does the example both ways, in one step 1000 000 × (3× 0.05) + 1000000, and in three steps.

Jo: I want you to look at this answer compared with our one-step answer. What do you notice about them?

S: They're different

[...]

Jo: Can anybody account for that difference?

S: what do you mean "account for"?

Jo: Why are they different, Una?

Una: ... Every time you get a different answer. You can't just go 1 million times [.15]

Jo: think of what she said. Each time we did a get a different answer, didn't we?

Ss: yes

Jo: Chukie?

Chukie: [...] If you do .05 times a million, you get 50 000; then you're gonna have to do it again; then you add a little bit to that 50 000 so you're not starting from 50 000, you're starting form a little more than 50 000. So, when you times by zero point five [0.05] again, you're gonna have a little bit more. No. I'm talking about ...

Jo: What he is saying here, when we do this, we're not multiplying just a million again, are we?

Ss: No.

Jo: We're multiplying a little more than a million.

Ss: Yeah

Jo: So, you're multiplying a little more this time, aren't you? And the next time, when you multiply a little bit, that gives us still...

S: A little bit more.

Jo: ... a little bit more than we had the first time, right?

Ss: Yes.

Jo: Instead of 50 000 we end up with 52 500. And we add that. We're gonna have a little bit more each time [...]

Jo writes on the overhead projector

    1 000 000 × 0.05 = 50 000
                   50 000
                   50 000

Jo: If we add, we were multiplying a million each time. This way we multiply by   1,000,000
     1,050,000
     1,102,500

Jo: [...] So you were multiplying more each time, right?

The group has spent so far more than 30 minutes discussing this problem. There seems to be consensus among the students that this is the proper method to use. There are however still some doubts. The cause for confusion in one students is a common misconception "multiplication makes bigger".

S: How do you multiply that? I don't get... I mean one million times .05 is smaller than what you started with?

Jo: Okay. And that's something that [...] you asked the other day, and we talked about that while you were absent. but let's take a quick look at it.

S1: Did you use your calculator?

S: But it was smaller than what you started out with!

Jo writes on the board, without solutions

$$\begin{array}{cccc} 10 & 10 & 10 & 10 \\ \times\ 1 & \times\ 2 & \times\ 8 & \times\ .5 \end{array}$$

Jo: If we are multiplying 10 × 1 what are we going to get?

S: Ten

Jo. [...] Do you all agree with me? Okay. Now let's take our same 10 and multiply it by .5. What are you getting? Where does the decimal go?

S: In the middle.

Jo: What is this?

S: Five point zero

Jo: That's 5

[...]

Jo: Right, we're only multiplying by half of a number here, aren't we?

S: We start with 10; we gotta have at least more than ten!

Jo: Do we?

S2: No

S: It's multiplying, it's multiplying. It's not?

Jo: We're saying if we have ten one time, we got ten, right?

S: Right.

Jo: What if we have ten one-half time?

S2: Five.

S: That's dividing though!

Jo: That's exactly the effect that we have, what we thought of as dividing, isn't it?

S: Yes, but there is a multiplication sign up there!

Jo: I know. If you have it a half time, you don't have ten, do you?

S: No, you've got it five times

Jo: That's exactly what we were doing.

S: So, 20 times 0.5 would be 10?

Jo: Yeah

S: Cool! Okay, I understand for that number we started ...

Altogether the class spent 47 minutes discussing one problem and its solution methods, including the clarification on why multiplying by a number that is smaller than 1 makes the product smaller. It is not common to find teachers willing to spend so much time on a single problem. In this excerpt of one of Jo's lessons, it is evident the richness of the mathematical ideas discussed, and the idea Jo is conveying through the discourse of what mathematics is about. Based on our multiple observations of Jo's teaching we can say that this example is fairly typical.

**Tools for enhancing discourse.**
Jo uses and encourages students to use a variety of tools to enhance the mathematical discourse in the classroom.

Computers, calculators and other technology
Jo has a very well defined posture with respect to the use of calculators. As she said during the interview, some students have trouble with the paper and pencil algorithms for the operations, "but that does not mean that they cannot understand how to solve problems and how to do other things in mathematics if they use a calculator as a tool."
Students in Jo's class have calculators available. In the factor game, students used calculators to verify factors of a number, and to add points for the game. Jo uses the blackboard and the overhead projector to facilitate communication with her students. Students use also the blackboard and the overhead projector to share their solutions and processes with other students.

Concrete materials used as models
Jo uses manipulative materials, although in the interview she stated that right now she has a problem because she does not have a lot, coming into a new classroom. She has a good supply of pattern blocks and she likes and uses those for fractions. She has some activities where they adapt to a lot of different things. She does not have Base 10 blocks or anything that is really good for place value. She would also like to have some geoboards.

Pictures, diagrams, tables and graphs
Several graphical representations of rational numbers were used: a whole divided into equal parts, a set of circles, rectangles or other figures divided into equal parts. Equivalent fractions were displayed systematically forming a table.

Invented and conventional terms and symbols
Conventional mathematical terms, like factor, multiple, remainder, quotient,

are introduced within a context, in a natural way, but not to the exclusion of other terms, and permitting the nonconventional use of those terms. In fact, students use unconventional terminology frequently. When talking about factors of 30, one student used the expression "multiply to" (analogous to "add to"). When talking about angles of tiles a student said that one angle "goes into" another (as in division of numbers 3 goes into 6).

Metaphors, analogies and stories
Jo uses realistic story problems. The meaning of special terms is explained. For example, in the case of the stories about fishing, terminology like "lures" were explained. She also uses whimsical or fantastic stories that are appealing to students.

Written hypotheses, explanations, and arguments
Interview: "I might ask them, 'what is a decimal?' and have them write for a certain length of time and write everything they can think of. And then we will discuss that generally within the class." "If I am walking around and I see some interesting thing about their writing, then we will discuss those as a way to develop the idea that we were working on."

## Conclusion
The fact that we have focused on the discourse in Jo's mathematics class does not mean, at all, that she does not also exemplify the other standards for teaching mathematics, such as giving her students worthwhile mathematical tasks, or creating an environment propitious for learning, and a continuous analysis of teaching and learning. Nor does it mean that she neglects at all the standards for the evaluation of the teaching of mathematics. Jo's teaching could be described using any of these as a thread. As we described the discourse in her classroom, it was also clear that she deals with the other standards for teaching mathematics, and with the standards for the evaluation of the teaching of mathematics. The mathematical discourse cannot be of high quality, as it is in Jo's classroom, unless there is also a high quality in the other standards.

Jo also focuses on the four standards that are common to all grades in the *Curriculum and Evaluation Standards* (NCTM, 1989): mathematics as problem solving, communication, reasoning, and mathematical connections. And with respect to the content, she teaches in concordance with the other content standards, although in this study we were focused on multiplicative structures and rational numbers.

Jo is not a teacher without problems, however. In the past, lack of understanding from parents and administrators of what she tries to accomplish in mathematics has been an issue. Among the barriers she meets when trying to teach the way she likes are having to cover the book to meet the district's regulations, limited copying capabilities of the school so that additional activities could not be done. Also, Jo has to deal with something that unfortunately too frequent in classrooms, misbehavior of some of the students. In most cases, she solves the problems herself, without affecting the flow of the discourse. In some extreme recurring cases, however, to solve some of these problems a teacher needs the support of the administration and parents. If that support is not there, a single student with an overt misbehavior can disrupt the discourse in the classroom and create an additional tension for the teacher. Her classroom is not particularly quiet, and she does not expect it to be, or wants it to be quiet. However, she mentions that discipline problems also interfere with the classroom discourse. She says that "four of the students require as much attention in this respect as the other 26", and she thinks that is not fair for the rest. She does not receive as much support as she would want from the administration. Although she is allowed to send students out, Jo says that "there is not really much teeth in it" (principals feel the pressure not to suspend students). And in some cases, there is no support from the parents. She describes the two most difficult cases. In one case, parents do not come to talk about the problem even after having been requested by the principal four times. In the other case, the parent blames all of the student problems on the school. In spite of the discipline problems, Jo says that this is a nice group to teach.

The mathematical discourse is not equally easy to orchestrate in different classrooms, even for the same extraordinary teacher. In the interview after the week-long observation Jo said, that this class was very verbal from the beginning, which had not been the case in other years with other classes, where students needed more encouragement and guidance to express and share their ideas in mathematics.

Jo is an example that the high level of discourse called by the *Standards* can be found in real life conditions, in ordinary public schools, orchestrated, promoted and enhanced by a teacher.

**References**

Ball, D. L. (1991). What's all this talk about discourse? Arithmetic Teacher, 39(3), 44-48.

Bishop, A. (1985). The social construction of meaning - a significant development for mathematics education? For the Learning of Mathematics, 5(1), 24-28.

Cazden, C. B. (1986). Classroom discourse, In M. C. Wittrock (Ed.), Handbook of Research on Teaching (3rd ed., p. 432-463). New York: Macmillan.

Cobb, P., Yackel, E., Wood, T. L. (1992). Interaction and learning in mathematics classrooms situations. Educational Studies in Mathematics, 23, 99-122.

Eisenhart, M. A. (1988). The ethnographic research tradition and mathematics education research. Journal for Research in Mathematics Education, 19, 99-114.

Fitzgerald, W., Winter, M. J., Lappan, G., Phillips, E. (1986). Factors and multiples. Menlo Park, CA: Addison Wesley.

Hiebert, J., & Carpenter, T. P. (1992). Learning and teaching with understanding. In D. A. Grouws, (Ed.), Handbook of Research on Mathematics Teaching and Learning (p. 65-97). New York: Macmillan.

Jungwirth, H. (1991). Interaction and gender-findings of a microethnographical approach to classroom discourse. Educational Studies in Mathematics, 22, 263-284.

Lampert, M. (1989). Choosing and using tools is classroom discourse. In J. Brophy (Ed.), Advances in research on teaching (p. 223-264). Greenwich, CT: JAI Press.

Lindow, J. A., Wilkinson, L. C., Peterson, P. L. (1985). Antecedents and consequences of verbal disagreements during small-group learning. Journal of Educational Psychology, 77, 658-667.

Miller, L. D. (1993). Making the connection with language. Arithmetic Teacher, 40, 311-316.

National Council of Teachers of Mathematics. (1989). Curriculum and evaluation standards for school mathematics. Reston, VA: NCTM.

National Council of Teachers of Mathematics (1991). Professional Standards for Teaching Mathematics. Reston, VA: NCTM.

Philipp, R., Flores, A., Sowder, J. & Schappelle, B. (1992). Reflective practitioners of mathematics teaching. (Manuscript submitted for publication)

Piaget, J. (1932). The language and thought of the child, 2nd ed. London: Routledge and Kegan Paul.

Pirie, S. E. B. (1988). Understanding: Instrumental, relational, intuitive, constructed, formalised...? How can we know. For the Learning of Mathematics, 8, 2-6.

Skemp, R. (1982). Communicating mathematics: Surface structures and deep structures. Visible Language, 16, 281-288.

Vygotsky, L. S. (1962). Thought and language. Cambridge, MA: MIT Press. (Original edition published in 1934)

Wood, T. L., & Yackel, E. (1990). The development of collaborative dialogue within small group interactions. In L. P. Steffe and T. Wood (Eds.), Transforming children's mathematics education (p. 244-252). Hillsdale, NJ: Lawrence Erlbaum.

Yackel, E., Cobb, P., Wood, T., Merkel, G. (1990). Experience, problem solving, and discourse as central aspects of constructivism. Arithmetic Teacher, 38(4), 34-35.

# 11 GEOMETRY OF NUMERIC ITERATIONS[11]

ABSTRACT: Simple numeric iterations can be represented by interesting geometrical patterns. Iterating different functions such as $x \to x^2 \pmod{p}$, and $x \to 2x \pmod{p-1}$ surprisingly generate essentially the same figures if $p$ is prime. The fact that two cyclic groups have the same structure is used to explain why the two diagrams have the same shape.

KEYWORDS: Numeric iterations, calculator, computer, isomorphism, cyclic group, logarithms, modular arithmetic.

**Introduction.**
This article presents some results in the area of discrete iteration, a developing field with applications in areas of mathematics, computer science, and the physical, life and social sciences. The result of iterating numerical functions such as $x \to x^2 \pmod{n}$ can be represented with geometrical patterns. These diagrams are many times beautiful. For example, Dewdney (1988) gives the diagrams for $n = 100$. The chains of numbers in the iteration process can be obtained with a calculator, or if the numbers are big, by using a computer program. For certain values of $n$, students can discover astonishing similarities in the diagrams for different functions. Here the diagrams corresponding to the iterations of the functions $x \to x^2 \pmod{p}$ and $x \to 2x \pmod{p-1}$ are explored. With these activities, students can see connections among mathematical topics such as geometry, algebra and computing, they can recognize equivalent examples of the same concept and relate procedures in them. They represent numeric problems using finite graphs. The activities

---

[11] Flores, A. (1994). Geometry of numeric iterations. *PRIMUS*, 4(1), 29-38.
Reprinted by permission of Taylor & Francis (http://www.tandfonline.com).

provide exploration opportunities in concrete examples of modular systems. It is an opportunity to see that seemingly different mathematical systems are essentially the same. Students can discover and prove elementary theorems using basic ideas of mathematical structures, such as groups.

### Activity 1. The iteration of $x \to x^2$ (mod p).

**Example 1.1.** Let $p$ be a prime number. Let us iterate the function $x \to x^2$ (mod $p$). That is, we take a whole number, square it, then reduce (mod $p$), and repeat the process with the result. We continue iterating until we get a number that has appeared already. We start with a new number and repeat the process until we exhaust all the mod $p$ numbers.

Let $p = 17$. The chains of numbers are

$0 \to 0$     $1 \to 1$     $2 \to 4 \to 16 \to 1$     $3 \to 9 \to 13 \to 16$
$5 \to 8 \to 13$     $6 \to 2$     $7 \to 15 \to 4$     $10 \to 15$
$11 \to 2$     $12 \to 8$    $14 \to 9$

We can represent these chains of numbers with a graph (see fig. 1). 1 is a big attractor, that is, once we arrive at 1, the iteration keeps giving the same value, and we eventually get to 1 by starting the iteration with any other number, except 0, which forms a cycle by itself. Notice that the numbers on the ends of adjacent branches, like 12 and 5, add to 17, and also for lower levels, for example, 8 + 9.

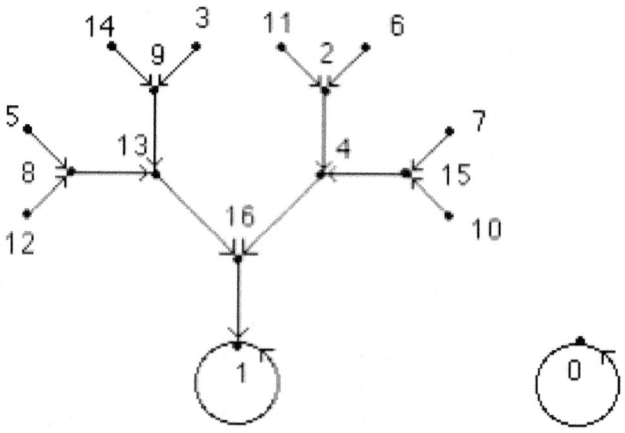

Figure 1. Iteration of $x \to x^2$ (mod 17)

**Example 1.2.** Let $p = 13$. By proceeding as before, the diagram now is formed by three parts. There are three attractors, where one is a cycle of two elements $(3 \leftrightarrow 9)$. Notice that this time the numbers on the ends of the branches add to 13.

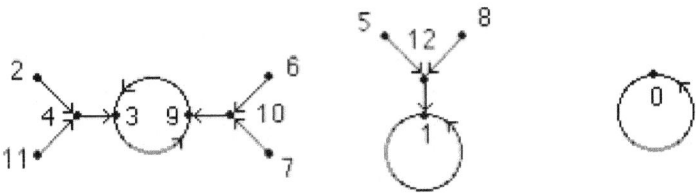

Figure 2. Iteration of $x \to x^2 \pmod{13}$

**1.3. Explaining one discovery.** In figure 1 the numbers on the extremes of the branches add to 17 (the module), and in figure 2 the numbers add to 13 (the module). We can see why this has to be the case, using clock arithmetic (modular arithmetic). By keeping the remainder when dividing by 17, we have essentially the same situation as in a clock, but with 17 "hours."

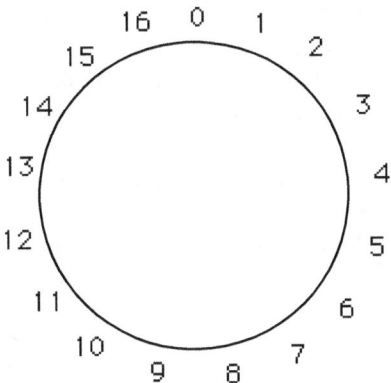

Figure 3. A seventeen-hour clock.

We can add numbers on the clock, for example $3 + 5 = 8$, $9 + 4 = 13$, $11 + 6 \equiv 0$. Numbers that add to 17 are additive inverses of each other in this arithmetic, for example $4 + 13 \equiv 0$, so that $13 \equiv -4$. Writing 13 as $-4$ makes it clear that 4 and 13 have the same square.

**Activity 2. The iteration of $x \to 2x \pmod{p-1}$**
**Example 2.1.** Let $p = 17$. We start with a number, multiply it by 2, reduce the result (mod 16). We continue iterating until we get a number that has already appeared, $1 \to 2 \to 4 \to 8 \to 0 \to 0$. We continue then with $3 \to 6 \to ...$ and so on, till we exhaust all mod 16 possibilities. Here is the corresponding graph.

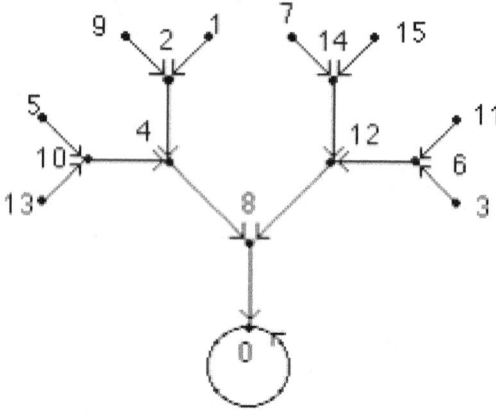

Figure 4. Iteration of $x \to 2x \pmod{16}$

Notice that all odd numbers are at the ends of the branches, at the next level are numbers of the form $2m$, where $m$ is odd, at the next level those of the form $2^2m$, etc. Notice that the difference between numbers at the end of adjacent branches is 8 (half the module).

**Example 2.2.** Let $p = 13$. We iterate the function $x \to 2x \pmod{12}$. The graph in this case is given in fig. 4. Notice that the difference between numbers at the end of the branches is 6 (half the module).

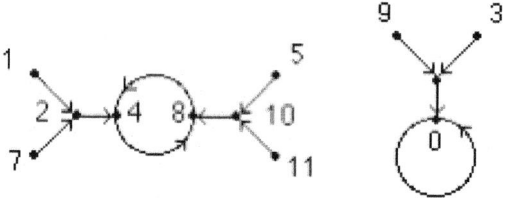

Figure 5. Iteration of $x \to 2x \pmod{12}$

**2.3. Explaining another discovery.** We noticed that in figures 4 and 5 the difference between numbers on the extremes of adjacent branches is half the module. Consider a "clock" with 16 hours. Numbers that differ by 8, differ by exactly half a circle. When multiplied by 2, the difference will be a whole circle, that is, the results will coincide.

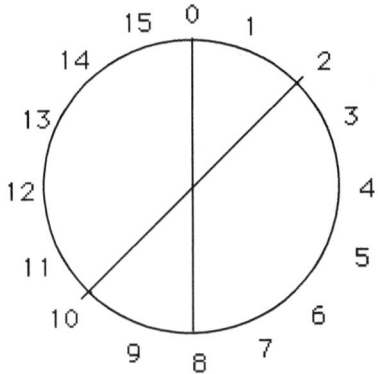

Figure 6. Multiplying by 2 (mod 16).

**Activity 3. An astonishing resemblance of diagrams.**
If we compare the graphs in 1.1 with the graph in 2.1, and that of 1.2 with that of 2.2, we see that graphs are identical for the iterations of $x \rightarrow x^2$ (mod $p$) and of $x \rightarrow 2x$ (mod $p - 1$), except for the extra cycle 0 which appears in 1.1 and 1.2. We can explain why the two diagrams are essentially the same by looking at the structure of $\{0, 1, 2, 3, ..., p - 1\}$, with multiplication (mod $p$) and that of $\{0, 1, 2, ..., p - 2\}$, with addition (mod $p - 1$).

**3.1. Groups with the same structure.** $\{1, 2, 3, ..., 16\}$ is a group under multiplication (mod 17), and $\{0, 1, 2, ..., 15\}$ is a group under addition (mod 16). They have the same number of elements, and furthermore, they have the same structure. We can make this explicit by using a generator in each group. 1 is a generator in $\{0, 1, 2, ..., 15\}$, +(mod 16), that is, adding it repeatedly we obtain 0, 1, 2, ..., 15. On the other hand, 3 is a generator in $\{1, 2, 3, ..., 16\}$, ×(mod 17); successive powers of 3, reduced (mod 17), give all the elements 1, 3, 9, 10, 13, 15, ..., 6. The one-to-one correspondence is given by table 1.

Table 1. A logarithmic table

| 0 | 1 | 2 | 3 | 4 | 5 | 6 | 7 | 8 | 9 | 10 | 11 | 12 | 13 | 14 | 15 |
|---|---|---|---|---|---|---|---|---|---|----|----|----|----|----|----|
| ↓ | ↓ | ↓ | ↓ | ↓ | ↓ | ↓ | ↓ | ↓ | ↓ | ↓ | ↓ | ↓ | ↓ | ↓ | ↓ |
| 1 | 3 | 9 | 10 | 13 | 5 | 15 | 11 | 16 | 14 | 8 | 7 | 4 | 12 | 2 | 6 |

Incidentally, this correspondence played a key role in the proof of young Gauss that the regular polygon with 17 sides is constructible with rule and compass (Gindikin, 1988). The numbers in the upper row are like the "logarithms" of the numbers in the lower row. To multiply two numbers in the lower row we simply add the corresponding numbers in the upper row and read the result below of the sum. This is obvious if we express the numbers in the lower row as powers of 3. Let $f$ denote the correspondence. Since $f$ preserves the operations in both groups, it is called an isomorphism, that is, $f(a \times b) = f(a) + f(b)$, $f(1) = 0$, $f(a^{-1}) = -f(a)$. In particular, if $y$ corresponds to $x$, the number corresponding to $x^2$ is $2y$ (see fig. 7).

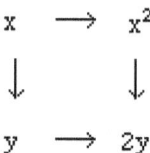

Figure 7. Corresponding elements and functions

This explains why the diagrams for the iterations of the functions $x \to x^2$ (mod $p$) and $x \to 2x$ (mod $p-1$) are identical. Corresponding elements have the same role in both diagrams.

Exercise. Give an explicit isomorphism between the groups $\{0, 1, 2, ..., 11\}$, $+$(mod 12) and $\{1, 2, 3, ..., 12\}$, $\times$(mod 13). Verify that corresponding elements have equivalent roles in both diagrams.

## What if $n$ is not prime?

What happens with the diagrams for $x \to x^2$ (mod $n$) and $x \to 2x$ (mod $n-1$) if $n$ is not prime? If $n$ is not prime, then $\{1, 2, 3, ..., n-1\}$, $\times$(mod $n$) is not a group. However, $\{0, 1, 2, ..., n-2\}$, $+$(mod $n-1$) is a group. The diagrams for the iterations of the two functions can be very different because there is not the same underlying structure.

**Example 3.2.** If $n = 9$, the diagram for $x \to 2x$ (mod 8) is a tree (see fig. 8), however, the form of the diagram for $x \to x^2$ (mod 9) is very different (see fig 9).

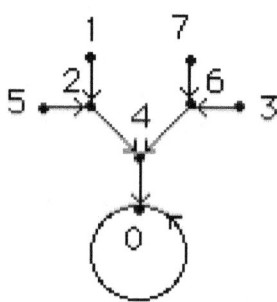

Figure 8. Iteration of $x \to 2x$ (mod 8)

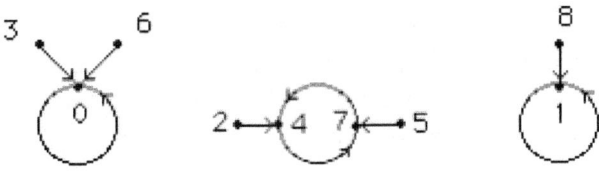

Figure 9. Iteration of $x \to x^2$ (mod 9)

Notice also that the difference of the numbers on adjacent branches is still half the module in the diagram of the iteration of $x \to 2x$ (mod 8), but that the sum of numbers on adjacent branches in the diagram of the iteration of $x \to x^2$ (mod 9) is not always 9. (Why?)

**Exercise 3.3.** If $n$ is odd, what will be the shape of the diagram of the iteration of $x \to 2x$ (mod $n$)?

## Activity 4. Extensions

If $p$ is prime, the diagrams of iterating $x \to x^3$ (mod $p$) and $x \to 3x$ (mod $p$-1) will be identical (except for the isolated 0 in mod $p$).

**Example 4.1.** Iterate the function $x \to x^3$ (mod 7). The diagram consists of 3 parts, the attractors are 1, 6, 0.

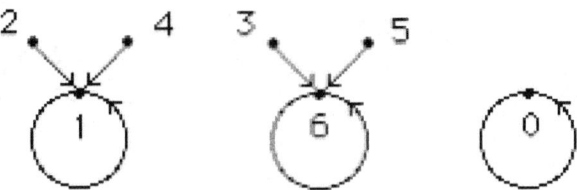

Figure 10. Iteration of $x \to x^3$ (mod 7)

**Example 4.2.** Iterate the function $x \to 3x$ (mod 6). The diagram has two parts; the attractors are 0 and 3.

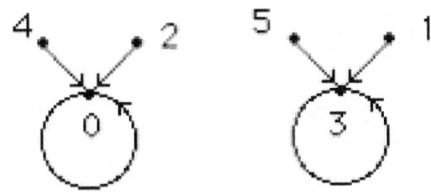

Figure 11. Iteration of $x \to 3x$ (mod 6)

In general, if $p$ is prime, the diagrams corresponding to the iterations of $x \to x^n$ (mod $p$) and of $x \to nx$ (mod $p - 1$) are identical. We can use one to describe the other. For example, what is the shape of the diagram $x \to x^2$ (mod $p$), if $p$ is a Fermat prime? It is easier to describe $x \to 2x$ (mod $p - 1$).

**Example 4.3.** Iterate the function $x \to 2x$ (mod $p - 1$), for $p = 257$, which is a Fermat prime, that is, of the form $p = 2^{2^k} + 1$

In this case the diagram is a tree. We show the big attractor 0, some of the intermediate nodes and some of the values at the end of the branches. All odd numbers are at the end of the branches, on each successive level the numbers are odd multiples of growing powers of 2. If $p - 1$ is a power of 2, the diagram for $x \to 2x$ (mod $p - 1$) is a binary tree. Therefore, the diagram for $x \to x^2$ (mod $p$), if $p$ is a Fermat prime, is also a binary tree.

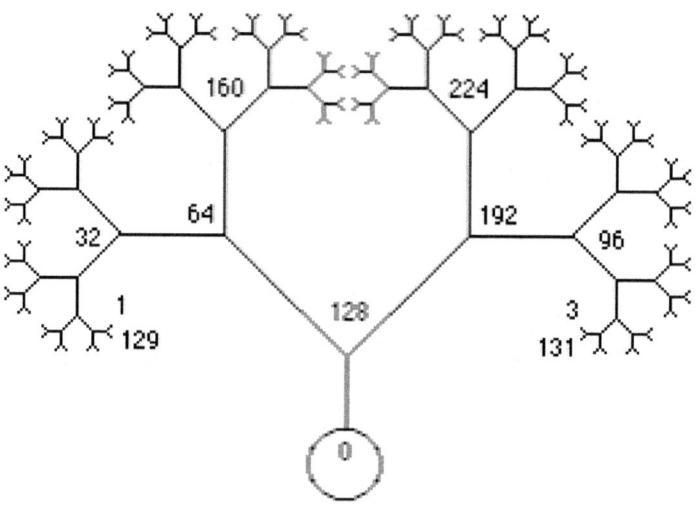

Figure 12. Iteration of $x \to 2x$ (mod 256)

**Conclusion.**

Representing numerical iterations by means of geometrical figures is a powerful tool in mathematics and is very attractive for students. It is surprising to find the same diagrams when iterating two different functions $x^2$ and $2x$. The idea of groups with the same structure helps to understand why the diagrams are identical. Students who do these activities have the opportunity of connecting ideas of three big fields in mathematics, relating numerical iterations with geometrical figures and with basic ideas in algebra. I have used these activities with mathematics majors and prospective mathematics teachers for grades 5-12. They enjoyed thoroughly the exploration projects based on these activities, especially representing geometrically numerical procedures, the discoveries of numeric patterns and relations in the diagrams, and the similarity of diagrams for different functions. The reviewer used discrete iterations with bright high school students and the students really got into it. Readers interested in applications of discrete iterations of the type discussed in this article may want to see Rosen (1993) where finding prime factors (up to $10^{15}$) of large integers is discussed using the Pollard rho method which involves the iteration $x_{k+1} = f(x_k)$ (mod $n$) where $f(x)$ is a polynomial with integer coefficients of degree greater than one (section 3.5, p. 156-159).

**References**
Dewdney, A. K. The armchair universe. N.Y.: Freeman, 1988.
Gindikin, S. G. Tales of mathematicians and physicists. Boston: Birkhauser, 1988.
Rosen, K. H. Elementary number theory and its applications (Third Edition). Addison Wesley, 1993.

BIOGRAPHICAL SKETCH
Alfinio Flores teaches mathematics "con ganas" to prospective teachers. He uses computers, calculators, and concrete materials to make mathematical abstractions meaningful to his students. He was trained as a mathematician at the National University of Mexico, and received a PhD degree in Mathematics Education from Ohio State University. At present, he enjoys the warm (human and climatic) environment at Arizona State University as Associate Professor.

# 12 CONNECTIONS[12]: A LOTTERY, A COMPUTER AND THE NUMBER e

Important mathematical constants, like $\pi$ and $e$ which are encountered first in very specific contexts, appear throughout the different branches of mathematics. It is surprising for students to find $\pi$, which they know as the ratio of the circumference to the diameter of a circle, in a probabilistic context, such as Buffon's needle problem (Hirsch, 1981). This article provides a link between Euler's constant $e$, and a probabilistic experiment such as a lottery drawing. This article will show how the number $e$, the base of natural logarithms, which students usually encounter in relation to compound interest problems, appears also in a probabilistic context. Establishing mathematical connection between different mathematical fields is one of the standards stressed throughout the K-12 mathematics curriculum in NCTM's *Curriculum and Evaluation Standards for School Mathematics* (1989).

The activities here described can be used as an extension activity after the constant $e$ is introduced. The article describes two experiments that could be conducted by students, one using a random table and one involving a computer simulation. The experimental probability results are then related to the theoretical probability and to the number $e$. Performing experiments to develop probabilistic intuitions and concepts before attacking theoretical probabilities, and stressing the relation between experimental and theoretical probability is recommended in the *Curriculum and Evaluation Standards for School Mathematics*. This approach has the advantage of getting the students involved

---

[12] Flores, A. (1993). Connections: A lottery, a computer and the number $e$. *Mathematics Teacher, 86,* 652-655. Copyright National Council of Teachers of Mathematics. Used by permission.

(by doing experiments) which is one of the recommendations of the *Professional Standards for Teaching Mathematics* (NCTM, 1991).

In the following discussion, we will refer to a lottery as a drawing where there is a single winning number. In a lottery with 100 numbers the probability for a player to win with a single draw is 1/100. Suppose there is a weekly lottery with 100 numbers. What would happen if a person would play in 100 such weekly lotteries? A very common and erroneous belief among lottery players is that if they play 100 times they will win at least once for sure, or almost for sure. The truth is that even when the probability of winning for each draw is 1/100, playing in 100 lotteries does not guarantee a win. The situation is very different from that of buying all 100 of the 100 tickets in one single lottery; in this case the player does have the winning ticket.

A simpler analogous case to our lottery is tossing a coin twice. The probability of guessing the outcome in a toss is 1/2, but the probability of guessing right at least once in two tosses is only 0.75, since the possible outcomes for the guesses are

| Right | Right |
| Right | Wrong |
| Wrong | Right |
| Wrong | Wrong |

and you are just as likely to guess correctly as incorrectly.

Listing all possible cases in a lottery would be cumbersome, but we can simulate the situation, first with a random number table and then with a computer program. Let us assume first that we have a lottery with ten numbers. Students can use a random table (statistics books usually have random tables, and one is provided in Table 1). Let them choose a number between 0 and 9. Students should look up their number in the first group of ten random numbers. They tally YES or NO accordingly if their number appeared at least once in the first group of ten numbers. This first group of ten numbers simulates ten drawings of the lottery. Students continue the process with more groups of ten random numbers, until they complete 50 groups.

NO _____  YES _____
How many NO's are there? _____. How many YES's? _____.
Ratio of NO's / Total _____. Ratio of YES's / Total _____.

Students can get an approximation to the probability of winning at least once in ten drawings in a lottery of ten numbers by computing the ratio of the number of YES's over the total of experiments, in this case 50. They can compute also the ratio of NO's.

For example, if their favorite number is 3, the total number of NO's using Table 1 would be 19, so that the proportion of NO's is 0.38
How many NO's are there? 19. How many YES's? 31.
Ratio of NO's / Total 0.38. Ratio of YES's / Total 0.62.

The ratios obtained this way represent the experimental probability based on a relatively small number (50) of simulations. Students who pick different numbers may get different experimental ratios. For example, for 2 as the favorite number, the ratio of NO's is 0.36 and of YES's is 0.64. Using a different random table would also give slightly different ratios. Students can learn from this experience that in random phenomena there is variability from one sample to another, from a group of simulations to another. For a small number of simulations, the experimental probabilities can vary considerably. One way to reduce the probability that the variation would be too big, is to use more simulations.

## Table 1

### Random number table

```
1 2 0 9 5 8 9 5 1 1
7 8 0 2 7 4 5 2 9 7
3 7 0 4 3 5 6 3 7 8
4 6 2 4 1 5 5 4 6 5
5 9 3 6 8 0 8 4 2 0
5 3 9 4 4 9 8 5 8 6
8 9 7 1 9 3 4 0 4 3
2 0 2 6 5 2 6 7 8 1
8 3 3 0 7 6 3 4 4 2
9 4 8 7 6 9 2 2 8 3
2 0 1 9 7 4 8 4 8 1
4 1 9 8 5 2 9 0 6 9
3 8 9 6 9 0 2 8 0 7
9 6 4 0 8 4 2 0 5 6
7 2 2 6 1 3 7 6 1 1
3 9 7 1 1 0 2 6 2 2
8 5 9 7 2 4 2 6 6 8
3 4 9 5 9 7 5 8 3 4
4 5 8 0 3 4 0 3 9 9
3 2 9 2 8 7 3 4 9 0
7 7 8 2 6 5 9 0 1 2
1 2 5 2 0 9 8 7 0 3
2 4 3 5 6 6 0 1 7 9
5 3 0 1 0 5 4 4 3 9
8 5 7 2 4 3 0 8 6 5
0 5 4 2 5 7 7 5 8 2
2 5 2 3 7 1 8 6 8 4
6 6 9 0 5 8 8 8 5 7
```

```
9 9 6 6 5 6 4 9 9 9
0 4 8 8 7 1 9 4 4 8
3 4 6 9 8 9 2 1 9 4
5 2 4 6 0 1 0 0 3 1
5 8 9 7 3 3 7 0 3 1
2 5 2 3 5 2 6 4 5 5
5 5 4 9 1 8 1 5 0 5
3 8 8 3 7 5 9 0 8 4
7 0 5 8 3 5 1 8 1 9
4 0 7 6 0 5 1 4 4 3
9 3 6 5 7 5 0 7 2 1
5 0 1 6 9 0 9 7 2 5
8 0 8 8 2 9 4 1 9 7
0 9 9 1 0 0 0 7 4 7
6 3 1 7 0 8 1 1 7 8
4 1 1 0 0 9 3 3 2 0
3 9 7 6 3 6 5 2 0 3
8 2 8 5 1 6 8 1 1 4
2 2 3 7 1 6 3 5 8 8
4 5 2 3 1 9 4 3 8 4
1 5 7 6 0 0 3 0 5 5
3 7 1 2 5 8 0 2 6 9
```

We can also simulate the situation with a computer. Program LOTTERY in this article, written in TrueBASIC can simulate the situation for a lottery of any number of tickets. When you run the program, you can choose the number of simulations, the total number N of tickets of the lottery, and your favorite number between 1 and N. For each simulation, the program tests whether you would get your favorite number at least once if you played the lottery N times.

### PROGRAM LOTTERY
REM LOTTERY TRUEBASIC
REM SIMULATES M SEQUENCES OF N DRAWINGS OF A LOTTERY WITH N NUMBERS
  PRINT "HOW MANY SIMULATIONS"
  INPUT M
  PRINT "HOW MANY NUMBERS HAS THE LOTTERY"
  INPUT N
  PRINT "CHOOSE YOUR FAVORITE NUMBER BETWEEN 1 AND ";N
  INPUT F
  RANDOMIZE

```
LET L = 0
FOR X = 1 TO M
LET S = 0
FOR Y = 1 TO N
LET R = 1 + INT(N*RND)
IF R = F THEN LET S = S + 1
NEXT Y
IF S > 0 THEN PRINT X; "WINNER",
IF S = 0 THEN PRINT X; "NO WINNER",
IF S = 0 THEN LET L = L + 1
NEXT X
PRINT "RATIO OF NO WINS TO TOTAL"; L/M
PRINT "RATIO OF WINS TO TOTAL"; (M- L)/M
END
```

Note: This program with minor alterations will run for other versions of BASIC. For example, for Apple II Basic, number all lines, delete RANDOMIZE, and use RND(1) instead of RND. For IBM PC Basic all you have to do is number the lines.

Let students run the program LOTTERY, or present it as class demonstration. Tell the computer to perform 50 simulations, where each simulation consists in performing 100 drawings for a lottery with 100 numbers. Choose a number between 1 and 100. In each drawing the computer picks one number at random, and after 100 drawings, the computer prints "WINNER" if your favorite number was chosen at least once, and prints "NO WINNER" if your favorite number was not drawn in any of the 100 drawings. It prints also the proportions of NO WIN's, and of WIN's. Table 2 is the outcome of running the program LOTTERY for 50 simulations, of 100 drawings each for a lottery of 100 numbers.

## Table 2
**Result of running the program LOTTERY**

HOW MANY SIMULATIONS? 50
HOW MANY NUMBERS HAS THE LOTTERY? 100
CHOOSE YOUR FAVORITE NUMBER BETWEEN 1 AND 100? 57
1 NO WINNER  2 WINNER   3 WINNER   4 WINNER   5 WINNER
6 WINNER   7 WINNER   8 NO WINNER  9 NO WINNER  10 WINNER
11 WINNER   12 WINNER   13 WINNER   14 WINNER   15 WINNER
16 WINNER   17 WINNER   18 NO WINNER   19 WINNER   20 WINNER
21 WINNER   22 WINNER   23 WINNER   24 NO WINNER   25 NO WINNER   26 NO WINNER  27 WINNER  28 NO WINNER  29 NO WINNER
30 NO WINNER  31 NO WINNER   32 WINNER   33 WINNER   34 WINNER   35 WINNER   36 WINNER   37 WINNER   38 NO WINNER

39 WINNER  40 WINNER  41 NO WINNER  42 WINNER  43 WINNER
44 WINNER  45 WINNER  46 NO WINNER  47 WINNER  48 NO WINNER
49 WINNER  50 NO WINNER
RATIO OF NO WINS TO TOTAL 0.32
RATIO OF WINS TO TOTAL 0.68

The ratio obtained based on the results of 50 simulations is the experimental probability for winning at least one time when playing 100 times in a lottery with 100 tickets. Run the program again, using 50 simulations and a lottery with 100 tickets. Run the program for more simulations, say 100, 200, but the same number of tickets.

We can run the program LOTTERY to simulate lotteries with more tickets. If we take a lottery with 10,000 numbers, what would be the probability of winning at least once if a player would play in 10,000 drawings? We can run the program to find out, but it will take much longer. Meanwhile (or instead), we can analyze another way to calculate the probability.

Since the probability of winning in a single drawing is 1/10,000, the probability of NOT winning is 0.9999. If a player draws twice in succession, the probability of not winning in any of the 2 drawings is 0.9999 × 0.9999, or $0.9999^2$. If the player draws three times, the probability of not winning in any of the 3 drawings is $0.9999^3$. If the player plays in 10,000 drawings, the probability of not winning in any of the drawings is $0.9999^{10000}$. Use a scientific calculator to compute this number. The result is 0.367861046... That is, the probability of winning at least in one drawing is only 1 - 0.367... = 0.632...

The number 0.367861046... turns out to be a very interesting number. Its multiplicative inverse is 2.718... which is very close to the number *e* = 2.718281828..., Euler's constant or the base of natural logarithms. How did this number *e* appear in this context of the lottery? Let us remember that *e* can be represented as

$$e = \lim_{n \to \infty} \left(1 + \frac{1}{n}\right)^n$$

or more generally,

$$e^x = \lim_{n \to \infty} \left(1 + \frac{x}{n}\right)^n$$

The expression that we are computing is $(1 - 1/10000)^{10000}$, or in general, for a lottery with *n* numbers, $(1 - 1/n)^n$. That is, in our case $x = -1$, therefore the limit of this expression when $n \to \infty$ is $e^{-1}$ or $1/e$.

The more numbers, N, a lottery has, the probability of a player of NOT winning in any of N drawings is closer to this limit number $1/e$. So, for lotteries with a large number N of tickets, the probability of a player who plays in N drawings of winning at least once is close to $1 - 1/e = 0.63212...$ Maybe knowing how low this probability is would make lottery players (with N numbers) realize the intrinsic difference between buying all N tickets in one drawing (certainty) and playing in N drawings.

Students are surprised at first by the fact that $e$ appears in a probabilistic context, but the more they learn mathematics, the more they perceive connections like this between different branches of mathematics as an example of the intrinsic unity of mathematics.

## References

Hirsch, Christian R. Probability and Pi. In Maletsky, Evan and Hirsch, Christian (Eds.) <u>Activities from the Mathematics Teacher</u>. National Council of Teachers of Mathematics, 1981. p. 111 - 114.

National Council of Teachers of Mathematics Commission on Standards for School Mathematics. <u>Curriculum and evaluation standards for school mathematics</u>. National Council of Teachers of Mathematics, 1989.

National Council of Teachers of Mathematics Commission on Teaching Standards for School Mathematics. <u>Professional standards for teaching mathematics</u>. National Council of Teachers of Mathematics, 1991.

# 13 THE SHADOWS OF MATHEMATICS[13]

*To Becky, Heidi, Sally, and Steve, who teach in the light*

My first glimpse of mathematics teaching in the United States was through conferences and workshops conducted by Donovan Johnson in Mexico. How lively and enjoyable the learning and teaching of mathematics could be! My vision was expanded when I became a member of the National Council of Teachers of Mathematics, and eagerly read the *Mathematics Teacher* and the *Arithmetic Teacher*, the yearbooks and other NCTM publications. What a wealth of ideas, what richness of approaches! Discovery learning, mathematics laboratories, games, activities with manipulatives, applications. Many ways to construct meanings, a lot of light in the learning and teaching of mathematics. I expected that in the United States the teaching of mathematics would reflect to a large extent this wealth.

However, I have found over the years that a large number of students in the United States have a very different picture of mathematics -- rote learning, meaningless procedures, unrelated topics, memorization of formulas. For them, learning mathematics is developing skills in symbolic manipulation of numbers and formulas, little understanding, and no fun. Year after year, these students have dealt only with the shadows of mathematics. Why is it that in spite of the leading role of a strong professional organization like NCTM, with a long tradition of publications and professional activities, in spite of the wealth of materials and opportunity to use them, many students have only seen the shadows of mathematics? If we want the reforms proposed in the *Curriculum and Evaluation Standards* and the *Professional standards for teaching mathematics* to take place, it is important to understand why

---

[13] Flores, A. (1993). The shadows of mathematics. *Arithmetic Teacher, 40*, 428-429. Copyright National Council of Teachers of Mathematics. Used by permission.

having access to ideas and methods is not enough.

Part of the answer is a "regression tendency" in many preservice and inservice teachers, and in students. They encounter exciting new ideas and methods to teach and learn mathematics, embrace them enthusiastically, but abandon them after a while. I saw student-teachers try the ideas and approaches from their methods courses. In some cases, however, the more time student-teachers spent in schools, the less they tried these ideas, the more they taught by the book. Teachers who take inservice courses try the ideas with enthusiasm, but after a time, many revert to the traditional teaching. Similarly, students stop trying to make sense, and study just to pass the test. Many just want to know what formula to use, many prefer to do pages of exercises from the book.

Plato's allegory of the cave may serve as a parallel. In this story, there is a group of people who from childhood live permanently in a deep underground cave. Their movements are constrained so that they can only see what is in front of them. Behind them there is a fire burning. All they see are the shadows of objects cast on the wall of the cave. They are used to the darkness and learn to recognize the shapes on the wall, and become very proficient. One person is taken outside and is temporarily blinded by the bright light, suffering pain. In spite of the great disorientation and fear, he is not driven back to the safe world of the cave. Gradually he gets accustomed to the sight of the upper world. He sees three-dimensional, live objects. He realizes that the shapes on the walls are just the shadows of these beautiful real objects. He runs inside to share his discovery. Plato knew very well the problems that such a seer would meet. The satisfied cave people would reject the suggestion that all their hard-earned skill of shadow recognition is inferior to another type of vision. The fact that the seer is no longer used to the dark, only makes his story less credible. He stumbles and is very clumsy. They laugh at this individual who cannot tell the shapes on the wall any more. Nobody believes him. This prophet could be rejected, even persecuted.

Unlike Plato's cave, the shadows of mathematics are not in a physical cave. It is not a building; it is not a place. Its nature is more subtle and therefore it is harder for people to realize they are in such a cave, and that the skills that they are learning are only the shadows of mathematics. The cave is a mind-set. It is part of a pervasive culture. As in the cave allegory, it is very difficult to be proficient both in the dark and in the bright light. Teachers who teach so that their students learn the real, live mathematics have to deal with standardized testing that measure preponderantly computational skills, not higher order thinking. They have to deal with textbooks with pages and pages of meaningless exercises, fragmented presentations, with a curriculum that has too many topics to be covered but is mandated. There is not time enough to let students construct their own meaning. Parents want their children to learn the same procedures and skills they learned. Teachers deal with students for whom mathematics has no meaning and who do not want to see a meaning, who expect learning to be boring. Students who think the relaxed atmosphere needed to discover meaning is an opportunity to disrupt. Many teachers who have seen the light are no longer confident teaching in the dark. They stumble, some fail. Some prefer to get used to the dark again. Many quit trying to teach real live mathematics and revert to safe practices, they teach computational skills so that their class gets good results in standardized tests, to please parents, or for some other reason. The students see only the shadows of mathematics.

However, over the years, I have also seen teachers who have preferred to make the continued effort to show their students real, live mathematics. It is a long way to the exit of the cave for many students, many resist being taken in that direction ("just tell me how to do it"). Once in the light, some students are blinded by the brightness and do not know what to do, they are irritated, they want to go back to the cave. It takes a long time for them to get used to the light. In spite of the difficulty, I have seen many magnificent, extraordinary, exemplary classroom teachers, having their students deal with real mathematics, ideas and concepts, not only with the shadows. The learning of their students is their best reward. However, it is important that we provide continuously the necessary support for them to carry out their labor, that we continue interacting and learning from them. These teachers will make the reforms proposed in the *Standards* possible. At the same time, achieving the quality of mathematics teaching and learning envisioned in the *Standards* by more teachers and students tomorrow, will be the best validation of their effort today.

References

National Council of Teachers of Mathematics. Curriculum and evaluation standards for school mathematics. Reston, Va.: The Council, 1989.

National Council of Teachers of Mathematics. Professional standards for teaching mathematics. Reston, Va.: The Council, 1991.

Plato. The Republic. In The Dialogues of Plato. Vol. 2. New York: National Library Co., 1900.

# 14 PYTHAGORAS MEETS VAN HIELE[14]

Mathematics is often taught using a spiral curriculum in the grades and junior/senior high school. The main idea of a spiral curriculum in mathematics is to view the same concepts several times with greater depth and understanding each time the topic is revisited. However, a curriculum intended to be spiral, can end as a curriculum that repeats itself, if we don't make sure that each visit is at a deeper level. A theoretical model such as the Van Hiele levels of development in geometrical thinking can serve as a guide to make sure that each time we revisit the topic we do it at a different level. According to Van Hiele (1986), there are five levels of development in geometry, which can be briefly described as follows:

### Level 1. Recognition

A student recognizes the shape as a whole. The student is not aware of specific parts or properties of that shape.

### Level 2. Analysis of properties

A student can analyze a figure, describe its parts and list its properties. Descriptions are used instead of definitions.

### Level 3. Informal deduction.

A student can understand the role of definitions, the relationships between figures; can order figures according to

---

[14] Flores, A. (1993). Pythagoras meets Van Hiele. *School Science and Mathematics*, *93*(3), 152-157. Used by permission of Wiley and School Science and Mathematics Association.

their characteristics; can deduce facts from previously accepted facts.

**Level 4. Axiomatic deduction.**

A student can understand the meaning of proof in the context of definitions, axioms and theorems.

**Level 5. Rigor.**

A student can understand the relationships between different axiomatic systems.

Several authors have documented that most high school students are not ready for formal proofs, and have emphasized the need for more informal geometry instruction prior to a formal axiomatic course in geometry (Hoffer, 1981; Senk, 1985; Shaughnessy & Burger, 1985).

The purpose of this article is
- to point to some preparatory activities for the Pythagorean theorem (level 1);
- give three different versions (Van Hiele levels 2, 3 and 4) of Euclid's proof of the Pythagorean theorem;
- give some generalizations of the theorem in the context of Euclidean geometry (level 4); and
- mention the Pythagorean relationship in other geometric systems (level 5).

Teachers reading this article should not think its content is intended to be presented to their students in a single course. This article covers too wide a range for any single course. Students need a certain educational maturity in going from one level to the next. The teacher, however, can select from it depending on the needs of the students.

The Pythagorean Theorem is one of the most important theorems in mathematics. Many formal proofs have been given (Loomis, 1968). Some of these proofs can also be adapted for students at different Van Hiele levels. Several proofs presented in a pleasant and dynamic way can be seen in a film (Apostol, 1988).

**LEVEL 1 Recognition.**

The Pythagorean relationship for particular cases can be presented with tangrams (see Figure 1), or the geoboard (see Figure 2), to familiarize students with the shapes that appear in the Pythagorean theorem.

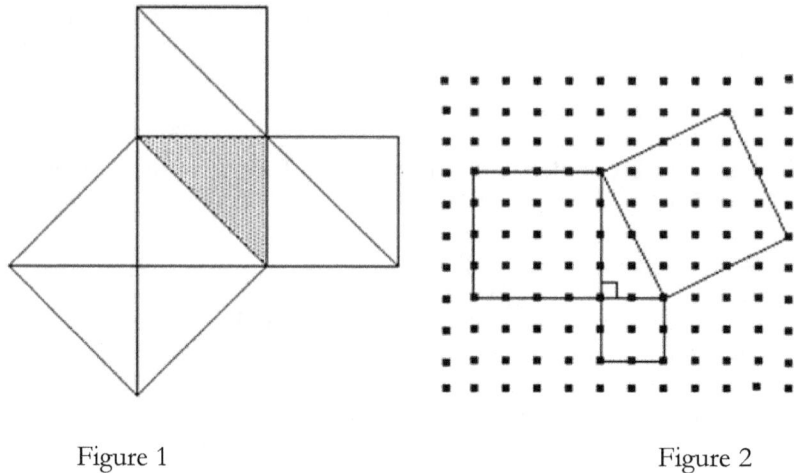

Figure 1                    Figure 2

## LEVEL 2 Analysis.

The purpose of this Pythagorean puzzle (Hall, 1974) is to make students aware of the different figures that are used later in the proof of the Pythagorean theorem, and how their areas relate to each other.

Paste the following pieces (Fig. 3) on cardboard and cut them out.

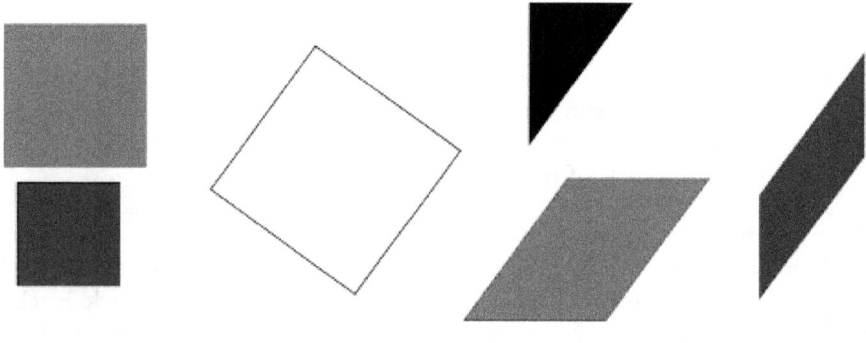

Figure 3

We will use the pieces to form the basic Figure 4 in various ways to see how their areas relate, and prove Pythagoras' theorem.

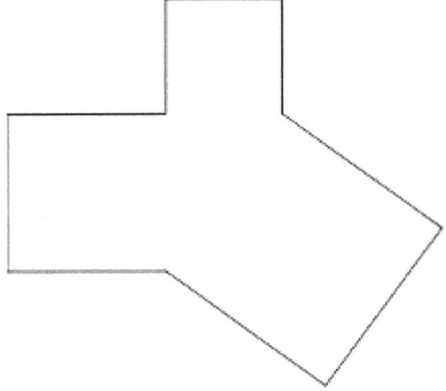

Figure 4

1) Use the triangle and the three squares to form the basic figure.

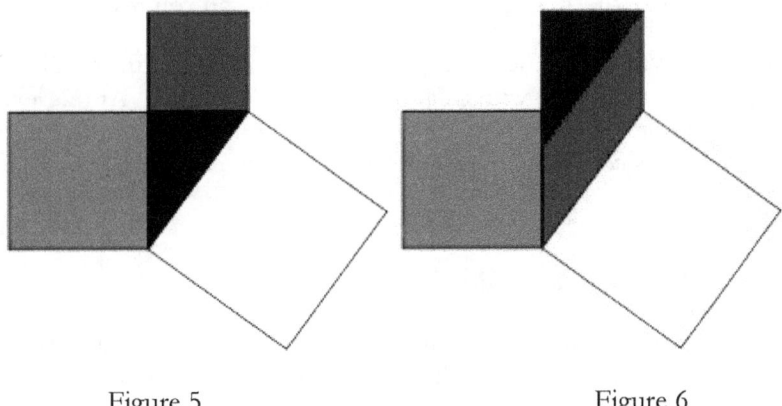

Figure 5                         Figure 6

2) With the triangle, the two bigger squares and the smaller parallelogram form the basic figure. What can you tell with respect to the areas of the parallelogram you have just used and the smaller square? Compare Fig. 5 with Fig. 6.

3) With the triangle, the biggest square, the smallest square and the bigger parallelogram form the basic figure. What can you say of the areas of the bigger parallelogram and the square you did not use? Compare Fig. 7 with Fig. 5

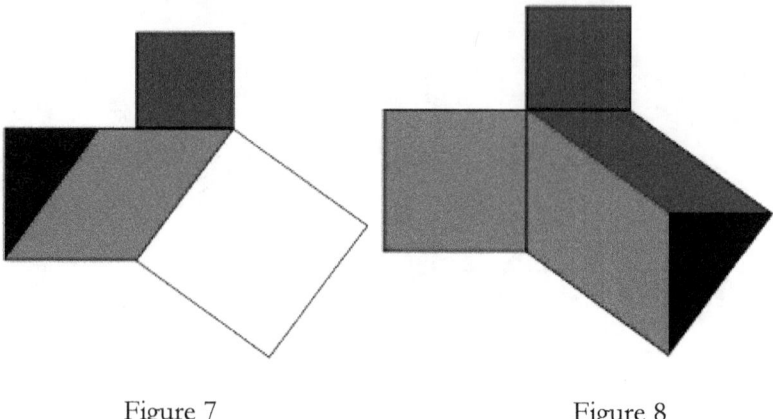

Figure 7  Figure 8

4) With the triangle, the two parallelograms and the two smaller squares form the figure. What can you say about the areas of the two parallelograms and the area of the big square? Compare Fig. 5 with Fig. 8.

Using the results of 2), 3) and 4), what conclusion can you reach?

Students at Van Hiele level 1 can be convinced of the truth of the Pythagorean relationship with an approach like this activity. At this level, no attempt is made to show from a mathematical point of view that the pieces of the puzzle, as constructed, fit exactly in the different positions.

## LEVEL 3 Informal deduction

Here is a proof that in a right triangle the square of a side is equal to the product of the hypotenuse and the projection of the side onto the hypotenuse (see Fig. 9)

$$b^2 = cq$$

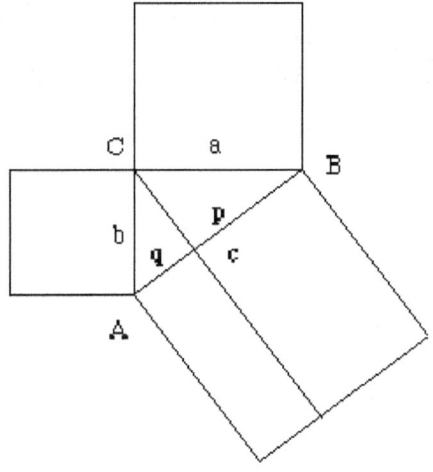

Figure 9

In the same way, it can be shown that $a^2 = cp$

The Pythagorean theorem can be obtained by simply adding the two results:

$$a^2 + b^2 = cq + cp = c(q + p) = c^2$$

Square ACSR has the same area as parallelogram ABR'R because they have the same base RA and the same height AC. (See Fig. 10)

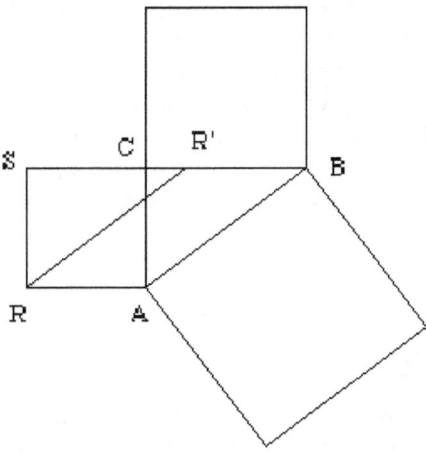

Figure 10

The parallelogram can be rotated 90° clockwise around A to parallelogram ATT'C (see Fig. 11). The segments RA and AC will coincide as will AB and AT. The angles BAR and TAC will also coincide. Parallelogram ATT'C has the same area as rectangle ATT"H, because they have the same base and the same height. Thus, the original square ACSR has the same area as rectangle ATT"H, namely $b^2$.

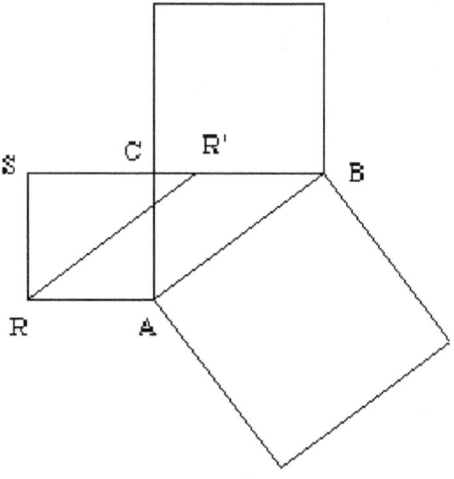

Figure 11

At this level, students are prepared to follow deductions using results previously accepted, or that do not need proof because they are evident. In contrast to the previous activity, students use some mathematical reasoning, to establish that the parallelogram fits in the different positions and that its area is equal to the square ACSR and to the rectangle ATT"H. However, the results are not deduced only from axioms and previously proven theorems. In the proof in this activity we used invariance properties of areas under rotations, which are intuitively obvious to students at this level, but that have not been proved.

**LEVEL 4 Axiomatic deduction.**

The following is essentially the proof of Proposition 47, Book 1 in Euclid's Elements (Heath, 1956). For this proof, Euclid uses only theorems that were previously deduced from the axioms and postulates.

Triangle ACR has the same area as triangle ABR because they have the same base RA and the same height AC (see Fig. 12).

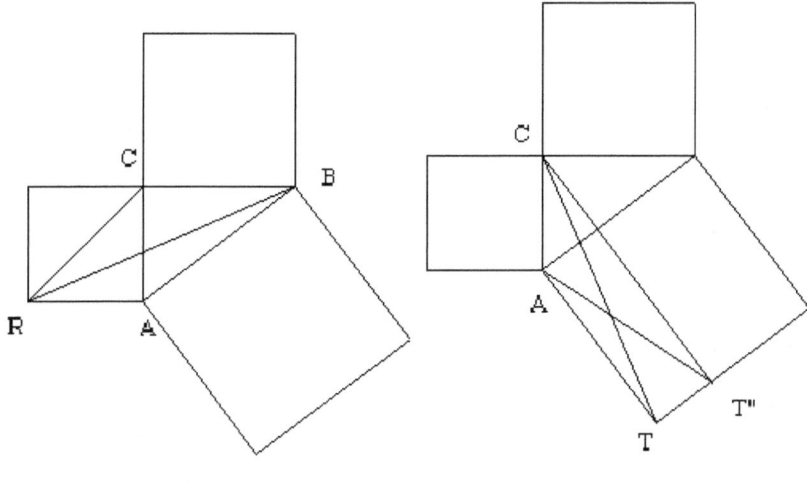

Figure 12                    Figure 13

Segments RA and AC are congruent to each other, as are segments AB and AT (see Fig. 13). Angles BAR and TAC are likewise congruent. Thus, triangle ABR is congruent to triangle CAT. Triangle CAT has the same area as triangle ATT", because they have the same base and the same height. Therefore, triangle ACR (see Fig. 12), which has half of the area of the square ACSR, has the same area as triangle ATT" which has half of the area of rectangle ATT"H. Thus, the rectangle ATT"H has the same area as square ACSR, that is, $b^2$.

Another proof can be given using similar triangles and proportions, an approach used in many high-school geometry books. However, a proof along these lines has to wait until the theory of similar triangles and proportions is developed. The advantage to give Euclid's proof is that we can prove and use the Pythagorean theorem much earlier.

The altitude at C of right triangle ABC partitions the original triangle into two right triangles AHC and CHB similar to the original (see Fig. 14).

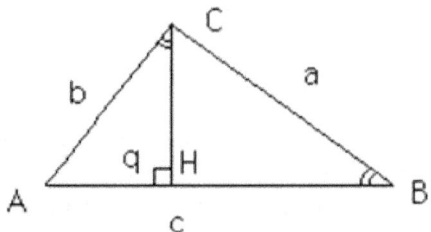

Figure 14

Since triangles ABC and AHC are similar, corresponding sides are proportional:

$$b/c = q/b$$

Multiplying both sides by $b$ and by $c$ we get

$$b^2 = cq$$

## Generalizations of Pythagoras

The areas of similar figures are proportional to the squares of their linear dimensions. The constant of proportionality $k$ depends on the shape of the figure. For example, the area of an equilateral triangle with side $c$ is $c^2 \times \sqrt{3}/4$, so that $k = \sqrt{3}/4$.

Since the two equations  (1) $b^2 + a^2 = c^2$

(2) $kb^2 + ka^2 = kc^2$

are equivalent, this means that the Pythagorean relationship holds not only for squares on the sides of a right angle, but for any kind of figures on the sides of the right triangle (see Fig. 15), as long as they are similar (Proposition 31, Book 6).

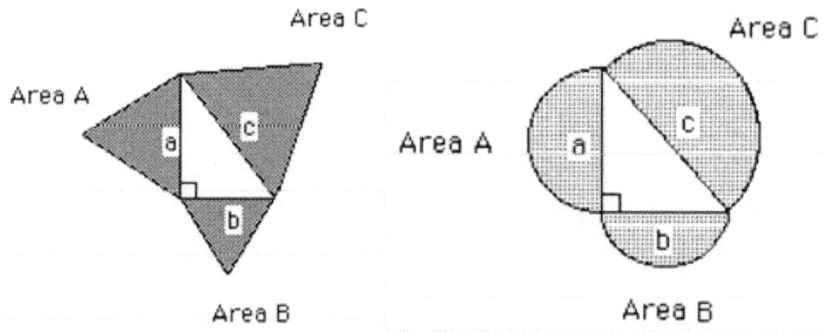

Figure 15

In order to prove equation (2), it is enough to prove that it holds for any specific value of $k$ ($k \neq 0$), that is, for a particular type of shape. We can view Figure 14 as a right triangle with three similar right triangles on its sides, realizing that the triangles are drawn to the inside instead of to the outside. The triangles on the side and base are AHC, and HBC, and the triangle on the hypotenuse is ABC itself. The sum of the areas of the two triangles on the base and the side is clearly equal to the area of the triangle on the hypotenuse.

**Pythagoras in space.**

The Pythagorean theorem can also be extended to three dimensions: In a right parallelepiped, the square of the length of the diagonal is the sum of the squares of the three sides.

$$d^2 = a^2 + b^2 + c^2$$

**The law of cosines**

In any triangle, $b^2 = a^2 + c^2 - 2\,a\,c \cos ß$, where ß is the angle between $a$ and $c$. The Pythagorean theorem is a special case when $a$ and $c$ are perpendicular, and $\cos ß = 0$.

**Pythagoras and trigonometry.**

The Pythagorean theorem is equivalent to the fundamental trigonometric equality

$$\sin^2 x + \cos^2 x = 1$$

This equation holds for any angle $x$, and not just for the angles that are possible in a right triangle.

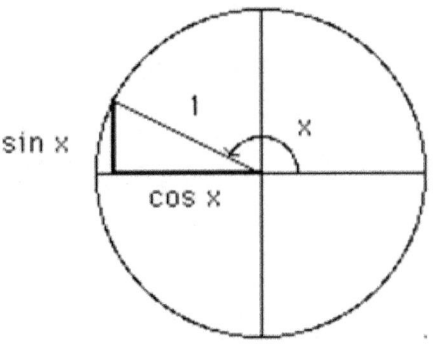

Figure 16

The previous four extensions of the Pythagorean theorem are in fact equivalent to it.

**LEVEL 5. Rigor.** Comparison with different axiomatic systems.

**Pythagoras and analytic geometry**

In a coordinate plane the distance d between two points $(x_1, x_2)$ and $(y_1, y_2)$ is usually defined by using the Pythagorean theorem as

$d = [(x_1 - y_1)^2 + (x_2 - y_2)^2]^{1/2}$     (Euclidean distance).

The Euclidean distance can be generalized for coordinate spaces with more dimensions.

However, distance in a coordinate plane can also be defined as

$d = |x_1 - y_1| + |x_2 - y_2|$     (taxi cab distance)

The locus of the points "equi-taxi-distant" from a given point is no longer a circle, and the locus of the points "equi-taxi-distant" from two points is in general not a straight line in this geometry (Smith, 1977). We do not have the Pythagorean theorem in taxi cab geometry.

**Pythagoras and vectors**

In $\mathbf{R}^2$ as a vector space, with the inner product of two vectors defined as usual

$(x_1, x_2) \cdot (y_1, y_2) = x_1 y_1 + x_2 y_2$

we have the relationship     $\mathbf{x} \cdot \mathbf{y} = |\mathbf{x}| |\mathbf{y}| \cos ß$

where ß is the angle between the vectors.

Reversing the process, angles between vectors can be defined in vector spaces that have an inner product. Perpendicular vectors satisfy $\mathbf{a} \cdot \mathbf{b} = 0$. Thus, if $\mathbf{a}$ and $\mathbf{b}$ are perpendicular,

$|\mathbf{a} + \mathbf{b}|^2 = (\mathbf{a} + \mathbf{b}) \cdot (\mathbf{a} + \mathbf{b}) = \mathbf{a} \cdot \mathbf{a} + 2 \mathbf{a} \cdot \mathbf{b} + \mathbf{b} \cdot \mathbf{b} = |\mathbf{a}|^2 + |\mathbf{b}|^2$

In the context of vector spaces with an inner product, the Pythagorean theorem is valid. However, vector spaces do not need to have an inner product defined.

**Pythagoras and spherical geometry**

In a non-Euclidean geometry, for example, spherical geometry, the Pythagorean relation for the sides of a right triangle does not hold. This can be seen in an isosceles triangle formed by two meridians and the equator, where actually two of the angles are right angles (see Fig. 17).

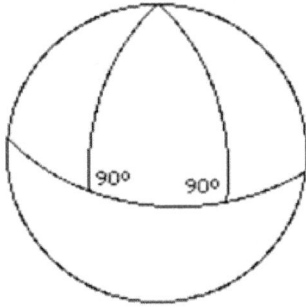

Figure 17

Mentioning these other contexts can make students aware that the central role that the Pythagorean theorem has in Euclidean geometry is maintained in some other systems, but not in all.

The teaching of mathematics can be done in a spiral curriculum if a guide such as the Van Hiele levels of development in geometry is kept in view. This article has looked at the Pythagorean theorem on all five of the levels. It was possible, hopefully it was enriching, and perhaps it was fun.

**References**

Apostol, T. *The theorem of Pythagoras*. National Council of Teachers of Mathematics, 1988.

Hall, G. D. A Pythagorean puzzle. In: *Teacher-made aids for elementary school mathematics: Readings from the Arithmetic Teacher*, NCTM, 1974.

Heath, T. L. *The thirteen books of Euclid's Elements*. 3 Vols. Dover, 1956.

Hoffer, A. Geometry is more than proof. *Mathematics Teacher*, 74(1), 1981, p. 11-18.

Loomis, E. S. The *Pythagorean proposition*. NCTM, 1968.

Senk, S. How well do students write geometry proofs? *Mathematics Teacher*, 78(6), 1985, p. 448-456.

Shaughnessy, J. M.; and Burger, W. F. Spadework prior to deduction in geometry. *Mathematics Teacher*, 78(6), 1985, p. 419-428.

Smith, S. A. Taxi distance. *Mathematics Teacher*, 70, 1977, p. 431 - 434.

Van Hiele, P. M. *Structure and insight: A theory of mathematics education*. Academic Press, 1986.

# 15 MATHEMATICS FOR BILINGUAL STUDENTS IN GRADES K - 3: EARLY INTERVENTION AND IDENTIFICATION OF GIFTED POTENTIAL[15]

**Abstract**: Mathematics activities to help identify students with gifted potential were conducted in Spanish in 21 bilingual classrooms grades K - 3. The activities emphasize higher order thinking skills in mathematics and were done by all students.

**Importance of early identification of gifted potential**
Early identification of giftedness provides the opportunity for early intervention. Research confirms that, unless nurtured, superior academic abilities will not develop. Intelligence and achievement test performance of truly academically superior children decline in an environment where they are perceived to have only average intelligence (Karnes, 1980). The earlier gifted children are identified, the sooner their intelligence can be nurtured and their talents developed. Project Excel was designed to identify potentially gifted bilingual students at an early age. For those identified, the program provides enriched opportunities aimed at the development of higher level thinking, and make them ready to be placed in the district's Gifted and Talented Education program at third grade. Excel is specifically geared towards bilingual students (Pérez, 1991); however, the principles on which this project is based can also be helpful for gifted programs for the early grades in general. Project Excel is based on the following principles:
- Gifted children can be found in all cultural and linguistic groups.
- Students may have special talents and skills that are valued in their own

---

[15] Flores, A., Pérez, R. I. (1993). Mathematics for gifted students in grades K- 3: Early intervention and identification of gifted potential. *SCOPE, 92*(3), 33-41.

culture, but often go unrecognized as gifted and talented in the school culture.
- Multiple assessment measures are needed to ensure appropriate screening.
- Different learning styles as well as different cultural values may inhibit student performance and subsequent identification of giftedness.
- Teachers need to be trained to recognize potential and to offer a curriculum that will evoke exceptional abilities.
- Potentially gifted students need an intensive program of enrichment.
- Parental involvement can have a significant impact on how well children use language skills, perform on tests, and behave in school.
- Mentors can provide positive role models for gifted students.

Identifying giftedness is not easy. Gifted children demonstrate different extraordinary proficiencies in a variety of intellectual and nonintellectual areas, such as reading, spoken languages, mathematics, artistic endeavors, etc. They do not do all of these things equally well. Neither do they show similarity to one another in personality, social functioning, or performance style. Gifted and talented persons are characterized not only by their above average ability and intelligence, but also by their creativity, and by their task commitment, that is, their intensive dedication to an activity (Ridge and Renzulli, 1981). Students, teachers, and parents are key elements in the complex task of identifying giftedness at an early age. Project Excel has one component for each one.

## Components of Project Excel
Project Excel has three main components: Student Readiness Component, the Teacher Training Component, and the Parent Involvement Component.

### Student Readiness Component
Project Excel focuses on developing students' strategies for higher order thinking in several areas, especially reading and oral language, but also in content areas such as science and mathematics. Students receive a qualitatively different curriculum from what is offered in regular classrooms. There is a change in quality in
    content - what they are taught;
    process - the way strategies and models are taught;
    product - results of student interaction; and
    learning environment - physical and psychological climate, which builds upon and extends the potential of high ability children.
One of the expected outcomes is that project students will be identified for a talent pool, and placed in the district's Gifted and Talented Education program at third grade.

Teacher Training Component.
Project primary teachers attend 20 inservice sessions each year. The process for the district's Gifted and Talented Education certification process includes seven categories:
1. Characteristics of the Gifted and Talented Student/Types.
2. Identification of the Gifted and Talented.
3. Theoretical Foundations of Giftedness.
4. Curriculum for gifted and talented learners.
5. Instructional techniques.
6. Parenting the gifted.
7. Professional growth.

One of the expected outcomes of the Teacher Training Component is that teachers will become effective in the early identification of gifted students so students can develop their potential and participate successfully in Gifted and Talented Education programs. A second objective is that they integrate higher level thinking strategies in their teaching. Before the students' activities that are described in this article were conducted in the classrooms, a workshop on the Pythagorean theorem using concrete materials was conducted with the teachers, to give them a broader perspective of the main ideas of the activities.

Parent Involvement Component
This component was designed to give parents the opportunity to generate ideas about giftedness and learn how to nurture gifted potential at home. Topics included are:
- An orientation to the Gifted and Talented Education program;
- An orientation to Project Excel;
- Characteristics of giftedness;
- Early childhood development and intellectual abilities;
- Multiple intelligences;
- Concept development;
- Resolution of conflict.

**Mathematics activities for gifted children**
Included in the Student Readiness Component, and the Teacher Training Component were activities to foster higher level thinking and creativity in mathematics. The rest of this article will focus on the mathematics activities for the students.

The activities for students were conducted in a series of visits to 21 classrooms in grades K to 3. The explanations of the activities and discussions of the mathematical ideas involved were done in Spanish, as one of the objectives of the project is to enhance higher-order thinking in the child's primary language. The rationale for the type of mathematical activities

that were designed and conducted are based on characteristics of mathematically gifted children, and on NCTM's recommendations for gifted students (1987). Mathematically gifted students show several of the following characteristics (House, 1987):
- Have a preference towards mathematics when presented with a choice of activities;
- Master typical content more quickly and at an earlier age than their classmates;
- Often skip steps in problem solving and may solve problems in unexpected ways;
- Are more willing and capable of doing problems abstractly; often prefer not to use concrete aids;
- Are successful at looking for patterns and relationships and attempt to explain them;
- Concentrate for long periods of time on a problem that they find interesting;
- Have exceptional mathematical reasoning ability and memory;
- Are more likely to see relationships between a new problem and problems previously solved; enjoy posing original problems;
- Are capable of more independent, self-directed activities;
- Enjoy the challenge of mathematical puzzles and games.

Good programs for the mathematically gifted pay attention to several essential components (House, 1987). Intuitive geometry activities were chosen to present important ideas and concepts and their interrelations, using concrete representations, and at the same time give gifted children the opportunity to explore a mathematical topic in depth. Activities are in concordance with NCTM's Position Statement on Early Childhood Mathematics Education (1991), and NCTM's Standards (1989, 1991). For the design of the mathematical activities the following aspects were taken into consideration:
- sound and significant mathematics - important geometry ideas such as area, relations between figures like triangles and squares;
- sound pedagogy - the set of activities are worthwhile mathematical tasks, students took an active part, discovering relations for themselves, and constructing their own meanings;
- higher order-thinking skills - based on their findings, students had to reach conclusions;
- applications and problem solving - activities were presented as problem to be solved;
- communications skills - findings were communicated orally and shared with the class;
- encouragement of creativity - many strategies could be used to solve the

problems;
- learning resources - simple cardboard models of squares and right isosceles triangles were used.

The activities were presented to all the children in the class. Thus, the activities had to be challenging enough for gifted children, but also had to be within the reach of average students. Higher-order thinking activities in mathematics are appropriate for all children (Chancellor, 1991).

Below is a summary of the activities done by students. Children in kindergarten did only the first four activities. A minimum of technical terms were used (square, triangle), and informal language was used during the presentation ("use the small triangle"). The number of activities was increased for higher grades and also the presentation was adjusted for each grade. More vocabulary was added, more precise description of the figures was given (for example, "a square has four equal sides and four equal angles"), and quantitative relations among the figures were made explicit (for example, "the area of the square is twice the area of the triangle"). Right isosceles triangles of two sizes, and a square cut out from cardboard were given to the children.

**Summary of the activities done by the children**

Using one cardboard square, and one or more small cardboard triangles (figure 1) students were asked to cover various shapes, such as other small triangles, small squares (figure 2), bigger triangles (figure 3), parallelograms (figure 4), bigger squares (figure 5). All of the figures appeared several times in the handouts, in different positions. Students were also asked to make comparisons between different figures. Verbal and written instructions, and the discussions of the activities were in Spanish. For a detailed description of the activities see Flores (1995).

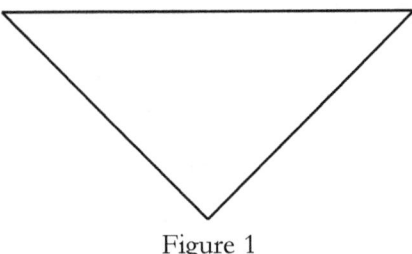

Figure 1

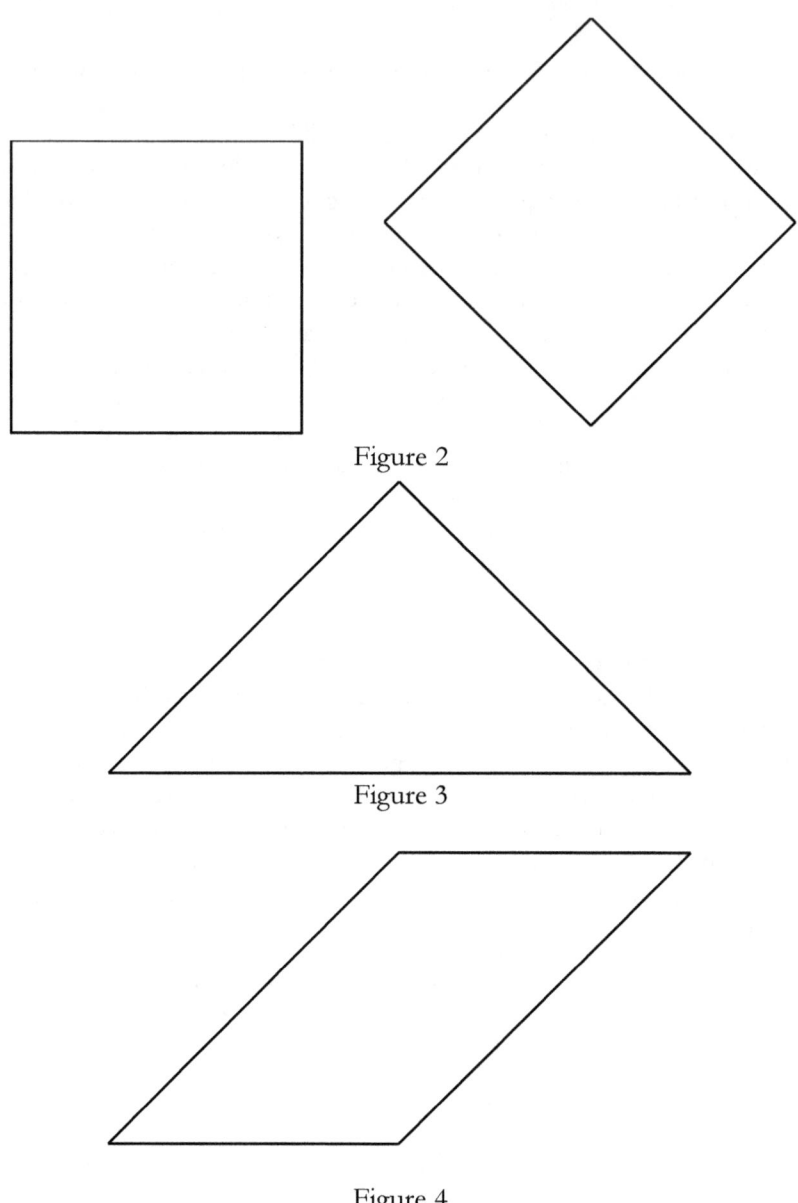

Figure 2

Figure 3

Figure 4

# TO CONNECT IS TO UNDERSTAND MATHEMATICS 4

Figure 5

**Mathematical discourse**

The concrete materials and activities provided the setting to develop mathematical discourse pertaining some geometrical facts. For example, students realized that all triangles that could be covered with the same cardboard triangle were equal, that position and orientation did not matter.

Some children called the second square in figure 2 a diamond, although after the activity they agreed that it was also a square, since it was equal to the cardboard square and to the other square.

When asked which figure would require more cardboard to be made (had more area), if the square or the big triangle, some children in second and third grade responded that the triangle, "because it is bigger." After pointing to them that both were formed by two small triangles, they expressed the fact that the two needed the same amount of cardboard (had the same area).

Students were asked to compare different shapes by using the smaller triangles, for example the bigger triangle and the square can both be covered with the same two small triangles; four little triangles are needed to form a big square (whose side is the diagonal of the smaller square), the same as for two of these squares.

**Observations.**

In kindergarten, children were familiar with the shape of a square. They were able to recognize and correctly show squares in their classrooms (calendar, boxes, frames, tiles, ceiling plafonds etc.) Only a few children pointed to some rectangles that were not squares.

All children did the activity of forming the square (in the two positions) with two triangles with relative ease. In this case matching equal angles or matching equal sides lead to the solution. Forming the big triangle took more time for some children, and some had to be helped, since they were using a strategy that was a dead end, matching corresponding angles from small and big triangles, so that there was no possibility to accommodate the second triangle (see fig 6). Owens (1992) has also noticed the importance of angles in spatial visualization tasks in her work with students in the early grades. Although the activities were not timed, it was apparent that forming the parallelogram with two triangles was easier for most children than forming the big triangle.

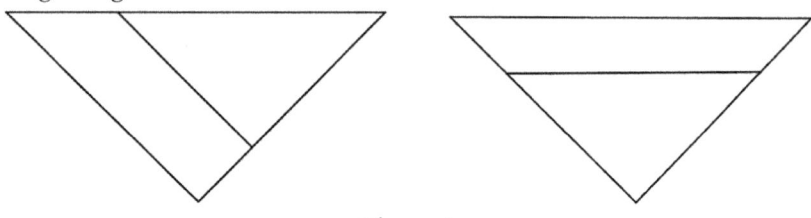

Figure 6

It was interesting to observe the different strategies used to form the same figure in different positions with the two triangles. Some children rotated the worksheet to get the new figure in a position similar to the figure they had solved first. Some children took the two small triangles together and translated and rotated both at the same time to fit into a new figure. Some children reassembled each figure from scratch, some of them always using the same method, some of them using trial and error each time.

Forming the big square with the four triangles was also challenging for most children, although some found the solution very quickly. Here again some of the students had to be helped since they insisted in matching equal (right) angles (see fig. 7), rather than trying to match equal sides to get the solution (see fig. 8).

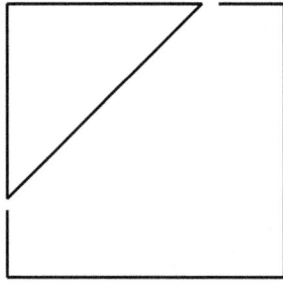

Figure 7

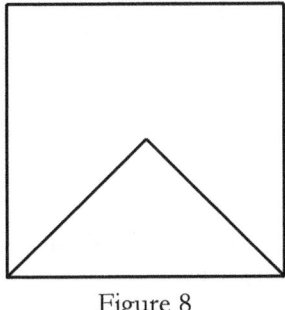

Figure 8

In a second grade, students were not given the frame in figure 5 when presented with the problem "to form a big square with four triangles." Although it took more time in general to solve the problem, the absence of the frame offered the opportunity for some children to find a different solution (see fig. 9)

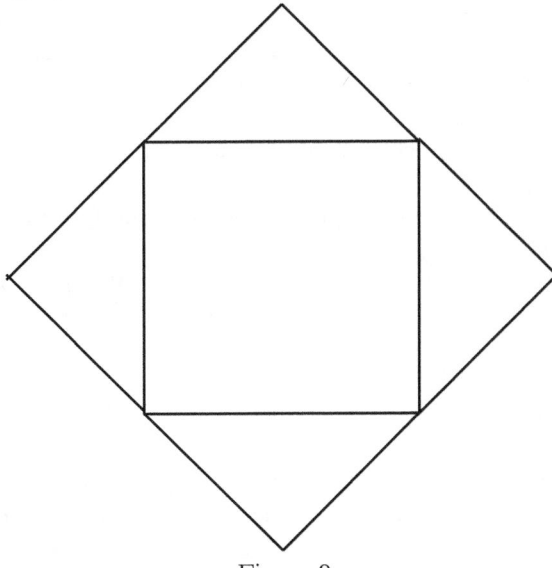

Figure 9

Students participated eagerly in the activities, some of them doing the activities ahead of the class on their own. Students liked the activities and were glad to keep the handouts.

Classroom teachers were present during the class presentation, and most teachers participated, helping students and answering individual questions. Teachers reacted positively to the activities presented. Some were surprised by the wide range of difficulty that children experienced when forming the figures with the small triangles, and also surprised by the variety of strategies used. It was a positive experience in geometry for students and teachers in the early grades, and an opportunity for all students, to develop and show their gifted potential.

**References**

Challenge: The mathematically able student [Special issue]. (1981). Arithmetic Teacher, 28 (6). (February).

Chancellor, Dinah. (1991). Higher-order thinking: a basic skill for everyone. Arithmetic Teacher, 38, (6) p.48 - 50.

Flores, A. (1995). Bilingual lessons in early-grades geometry. *Teaching Children Mathematics, 1*(7), 420 - 424.

Gifted students [Special issue]. (1983). Mathematics Teacher, 76 (4) (April)

Karnes, M. B. (1980). Rationale for early identification and programming for the gifted and components of an exemplary program. Paper presented at the American Educational Research Association, Boston, April.

House, P. A. (Ed.). (1987). Providing opportunities for the mathematically gifted, K-12. National Council of Teachers of Mathematics.

National Council of Teachers of Mathematics. (1991). Early childhood mathematics education. In 1991-92 Handbook. NCTM Goals, leaders, and positions. National Council of Teachers of Mathematics. p.16.

National Council of Teachers of Mathematics. (1987). Provisions for mathematically talented and gifted students. NCTM News Bulletin, 23 (January), 7.

National Council of Teachers of Mathematics. (1989). Curriculum and Evaluation Standards for School Mathematics. National Council of Teachers of Mathematics.

National Council of Teachers of Mathematics. (1991). Professional Standards for Teaching Mathematics. National Council of Teachers of Mathematics.

Owens, K. (1992, August). Young children's spatial processing and the development of spatio-mathematical reasoning. Paper presented at the 7th International Congress on Mathematical Education, Quebec, Canada.

Pérez, R. I. (1991). Project Excel. San Diego City Schools.

Ridge, H. L., and Renzulli, J. S. (1981). Teaching mathematics to the talented and gifted. In V. J. Glennon (Ed.), The mathematical education of exceptional children and youth (p. 191-266). National Council of Teachers of Mathematics.

Wilmot, B., and Thornton, C. A. (1989). Mathematics teaching and learning: Meeting the needs of special learners. In P. R. Trafton (Ed.), New directions for elementary school mathematics (p. 212-222). National Council of Teachers of Mathematics, 1989.

# 16 A GEOMETRICAL APPROACH TO MATHEMATICAL INDUCTION: PROOFS THAT EXPLAIN[16]

Lately, mathematics educators have placed a greater role on the concept of proof as a convincing argument, as a way to make proof a means of communication. Hanna (1990) suggests that whenever possible, we should give students proofs that explain rather than proofs that only prove.

Proofs that prove and proofs that explain are both valid proofs. A proof that explains, in addition, "must provide a rationale based upon the mathematical ideas involved, the mathematical properties that cause the asserted theorem to be true." (Hanna, 1990, p. 9)

Mathematical induction and proofs without words

Many times, mathematical induction is perceived by students as a method that proves a numerical formula, but that does not give them essential reasons of why the formula is so. In this paper, we present some examples that offer students an insight into why the formulas are true. At the same time these examples provide insight into the two parts of the method of proving by mathematical induction: the initial step and the inductive step. Proofs by mathematical induction can also be proofs that explain.

Figure 1a not only suggest the result that the sum of odd numbers is a square, but also how to extend the square into another square by adding the next odd

---

[16] Flores, A. A geometrical approach to mathematical induction: Proofs that explain. *PRIMUS*, 2(4) 393-400. Reprinted by permission of Taylor & Francis (http://www.tandfonline.com).

number (see fig. 1b), thus showing that $1 + 3 + 5 + \ldots + (2n - 1) = n^2$.

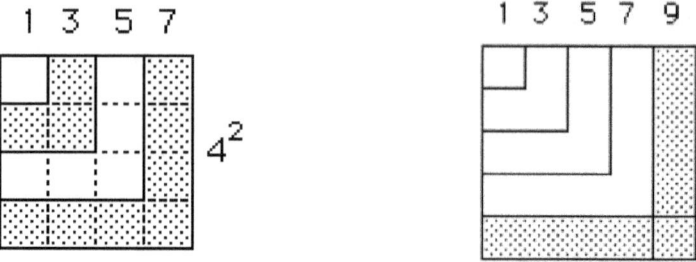

Figure 1a                    Figure 1b

In the same way, some diagrams that are published as "proofs without words" (for example, Cupillari, 1989; Nelsen, 1990; Zerger, 1990), or that are described as "proof by looking" (Wells, 1991), suggest the validity of numerical formulas. These diagrams have the advantage of being proofs that explain, giving students some insight into how these formulas can be derived. Some may object that such diagrams are not a proof, since they are only drawn for a particular number. However, an essential characteristic of such diagrams is that they suggest how we can form a diagram for the case $n + 1$, if we assume we have such a diagram for a value $n$. Such diagrams can thus be used to introduce students to proofs by mathematical induction, in a way that the proof both explains and proves.

I have used these examples in a course on problem solving and proofs, which is a transition course between the calculus sequence and the more formal analysis and modern algebra. Students enjoyed the alternative approach to induction. Many students who used to just crank out algebraically the case for $k + 1$ of a formula by manipulating the formula for the case $k$, had really to think through the meaning of the inductive step. Students who were used to think only in symbolic terms developed the ability to think about formulas in geometrical terms, although it was hard for some of them at the beginning. In general, students were able to use the diagrams to prove the formulas by mathematical induction and to solve the examples and exercises provided below, showing insight of the process.

EXAMPLES

1) The sum of odd numbers $1 + 3 + 5 + \ldots + (2n - 1)$

Figure 1a shows that $1 + 3 + 5 + 7 = 4^2$. We could easily extend the square by adding 9 (the next odd number) small squares to form a square of side 5, as shown in Figure 1b, and so on. However, to prove that $1 + 3 + 5 + \ldots +$

$(2n - 1) = n^2$ is valid for any number, a few particular examples are not enough. However, we can extend the diagram for the case $n + 1$, assuming we have a diagram for $n$ as suggested by fig 1b. In general, if we assume that the sum of the first $n$ odd numbers is a square of side $n$, we can show that we can add the next odd number $(2n + 1)$, represented as two rectangles of area $n$, and a unit square, and get a square of side $n + 1$.

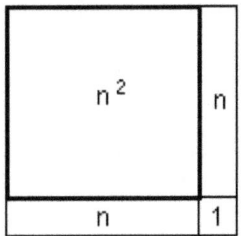

Figure 2

Figure 1a shows that the formula holds for an initial value of $n$, Figure 2 shows the inductive step, to prove the case $n + 1$ assuming the case $n$. By the principle of mathematical induction, the formula holds for all natural numbers.

## 2) Sum of cubes of the first $n$ natural numbers $1^3 + 2^3 + 3^3 + \ldots + n^3$

The triangle in figure 3 has a base of length $(4 \times 5)$ and height $(1 + 2 + 3 + 4)$ On the other hand, its area is equal to $1 \times 1^2 + 2 \times 2^2 + 3 \times 3^2 + 4 \times 4^2 = (1^3 + 2^3 + 3^3 + 4^3)$. Therefore,

$$(1^3 + 2^3 + 3^3 + 4^3) = (4 \times 5) \times (1 + 2 + 3 + 4)/2.$$

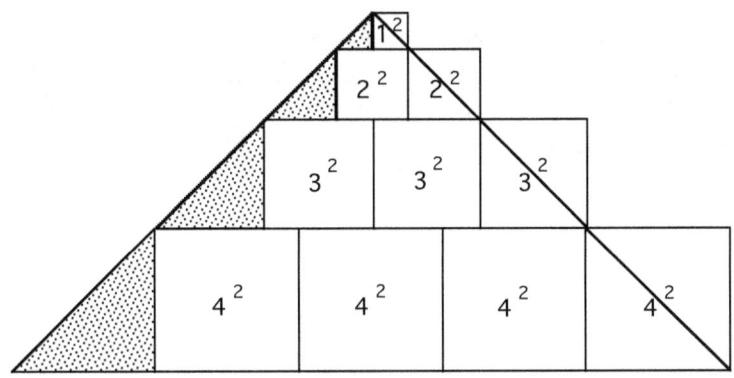

Figure 3

To prove that in general $(1^3 + 2^3 + 3^3 + \ldots + n^3) = n(n+1) \times (1 + 2 + 3 + \ldots + n)/2$ we need to prove the result for $n+1$ assuming the result for $n$.

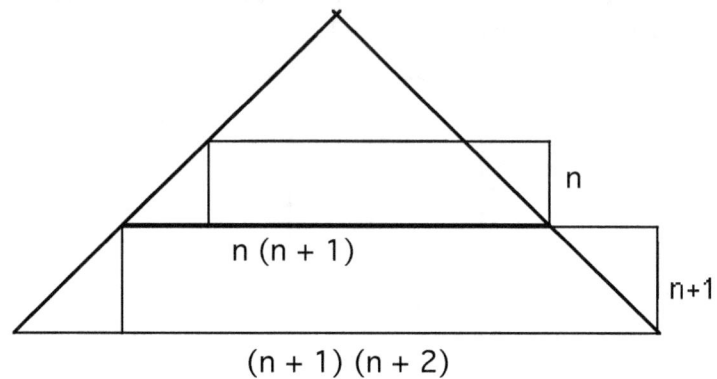

Figure 4

For case $n$, the base of the triangle is $n(n+1)$ and its height is $(1 + 2 + 3 + \ldots + n)$. Adding $(n+1)^3$ in the form of a strip of $(n+1)$ squares of area $(n+1)^2$, and cutting one of the squares along the diagonal and re-accommodating one piece we can extend the triangle to one of base $(n+2) \times (n+1)$ and height $[1 + 2 + \ldots + n + (n+1)]$ which is what we wanted to prove.


EXERCISES

1) The sum of the first $n$ natural numbers $1 + 2 + 3 + 4 + \ldots + n$. The rectangle in figure 5 has an area of $4 \times 5$, and on the other hand of $2(1 + 2 + 3 + 4)$. To prove that in general $1 + 2 + 3 + 4 + \ldots + n = n(n+1)/2$, assume that you have a diagram for case $n$ of this formula and extend the diagram to the case $n+1$.

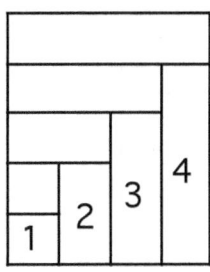

Figure 5

2) The sum of the squares of the first $n$ natural numbers. The following diagram suggests a way to calculate $1^2 + 2^2 + 3^2 + \ldots + n^2$ (Baron, 1969). Assume you have a diagram for the case $n$. Show that you can extend the diagram for the case $n + 1$. (This geometrical method was used by Ibn-Al-Haitham (Alhazen) about one thousand years ago. He used this method too to obtain the sums $\Sigma n^3$ and $\Sigma n^4$ and it can be used for sums of other powers of natural numbers.)

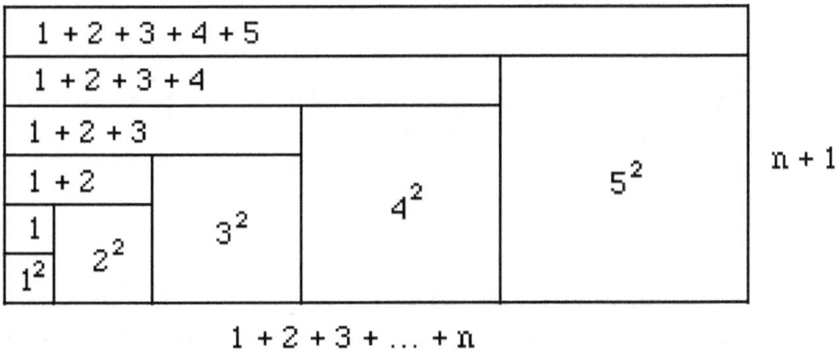

Figure 6

2) Sums of triangular numbers. Let $T_n = 1 + 2 + 3 + \ldots + n$. The following diagram (Zerger, 1990) suggests that $3(T_1 + T_2 + \ldots + T_n) = (n+2)(1 + 2 + \ldots + n)$.

Assume you have a similar diagram for $n$. Show how you can extend such a diagram for the case $(n + 1)$.

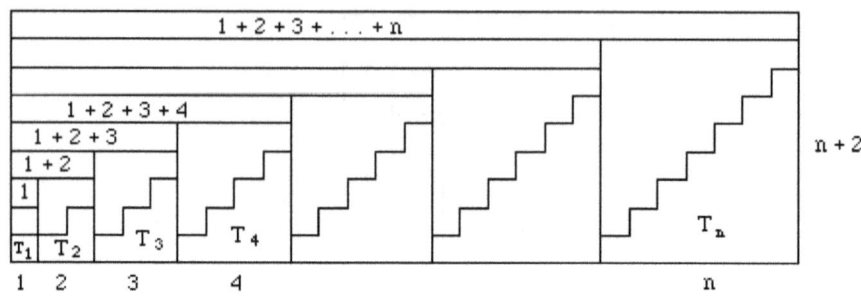

Figure 7

4) The following diagram (Wells, 1991) suggests that $8T_n + 1 = (2n + 1)^2$

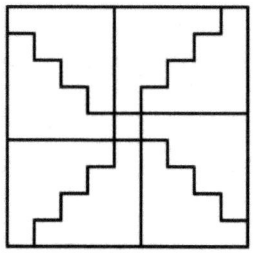

Figure 8

Assume that you have a diagram corresponding to $n$. Show that you can extend it to a diagram corresponding to $(n + 1)$.

5) Sums of cubes revisited.

a) Figure 9 (Wells, 1991) shows one $1 \times 1$ square, two $2 \times 2$ squares, three $3 \times 3$ squares and so on. The area of the square is therefore $1^3 + 2^3 + 3^3 + 4^3 + 5^3$. Notice that there is one $3 \times 3$ square on the corner and $(3-1)/2$ squares of size $3 \times 3$ on each side, one $4 \times 4$ square on the corner and $(4-1)/2$ four by four squares on each side.

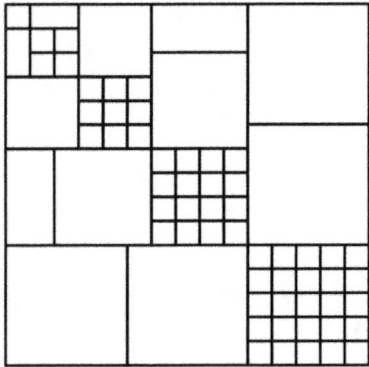

Figure 9

Show that if we have a square with sides $(1 + 2 + 3 \ldots + n)$ where the last set has $n$ squares of size $n \times n$, then we can extend the pattern into another square with sides $(1 + 2 + \ldots + (n + 1))$ by adding $(n + 1)$ squares of size $(n + 1) \times (n + 1)$. Illustrate for $n$ even and $n$ odd. (Hint: divide the set of $n$ squares $n \times n$ in three sets, the square on the corner, and two sets with $(n - 1)/2$ squares on each side. To fit the $n + 1$ squares $(n + 1) \times (n + 1)$ forming an extended square, we need that $(n + 1) \times n/2 = (n - 1)/2 \times n + n$.)

b) Figure 10 (Cupillari, 1989) shows that $4(1^3 + 2^3 + 3^3 + 4^3) = (4^2 + 4)^2$. To prove that $4(1^3 + 2^3 + 3^3 + \ldots + n^3) = (n^2 + n)^2$ we need to show that we can prove the case for $n + 1$ assuming we have the case for $n$.

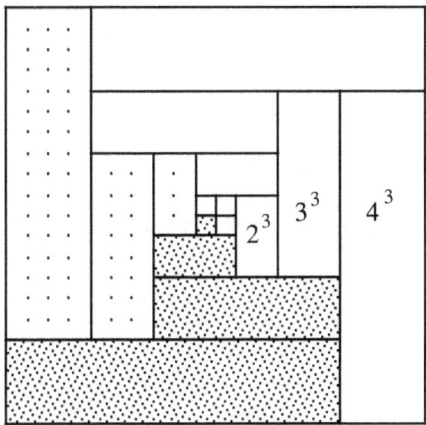

Figure 10

For the case $n$, the side of the bold square in Figure 11 is $n^2 + n$, and the outer rectangles are of size $n$ by $n^2$, so that their area is $n^3$. We want to show that we can add four $(n + 1)^2 \times (n + 1)$ rectangles to form a square with sides $(n + 1)^2 + (n + 1)$. This is possible because $n + n^2 + (n + 1)$ is precisely $(n + 1)^2$.

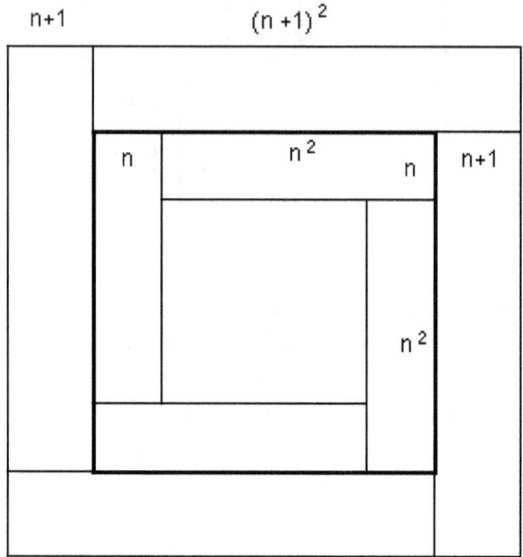

Figure 11

5) Prove that the sum of triangular number is the sum of squares. That is, if $T_k = 1 + 2 + 3 + ... + k$, then

$$T_1 + T_2 + T_3 + ... + T_{2n-1} = 1^2 + 3^2 + 5^2 + ... + (2n-1)^2$$

$$T_1 + T_2 + T_3 + ... + T_{2n} = 2^2 + 4^2 + 6^2 + ... + (2n)^2$$

## TO CONNECT IS TO UNDERSTAND MATHEMATICS 4

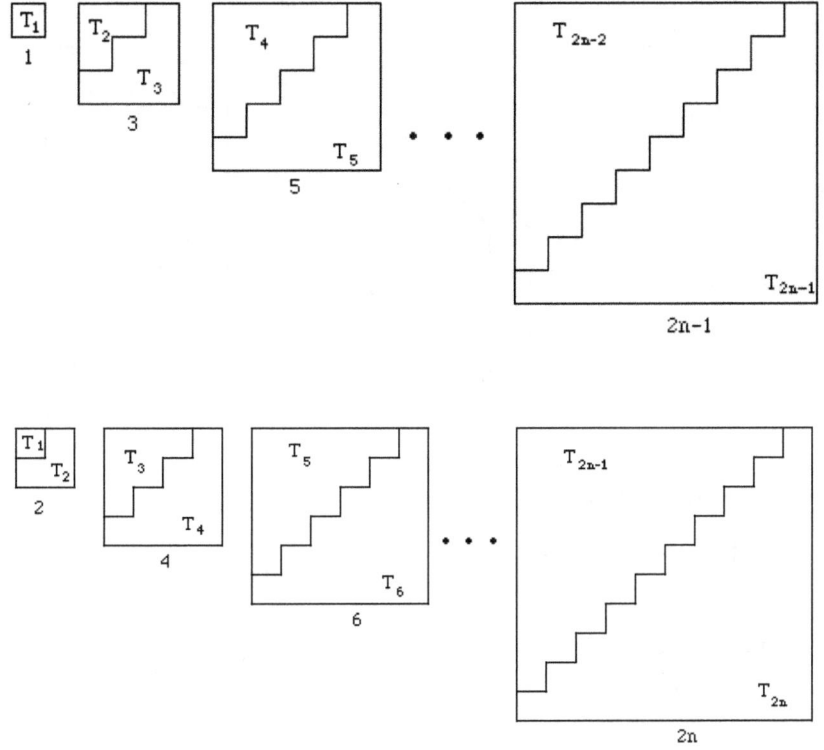

Figure 12

6) $(1\times2) + (2\times3) + (3\times4) + \ldots + (n-1)n = (n-1)n(n+1)/3$

The following diagram (Wu, 1989) shows the initial step and the inductive step in the proof of the statement.

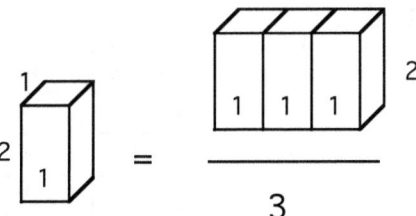

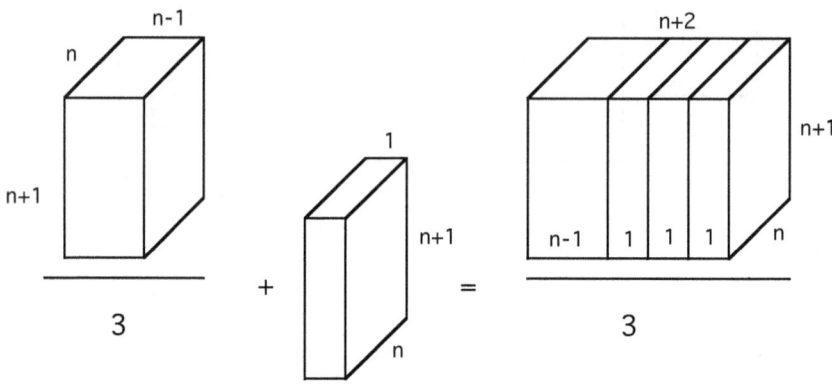

Figure 13

Sums of triangular numbers again

Let $T_k = 1 + 2 + 3 + \ldots + k$

$$(n+2)T_n = 3\sum_{k=1}^{n} T_k = nT_{n+1}$$

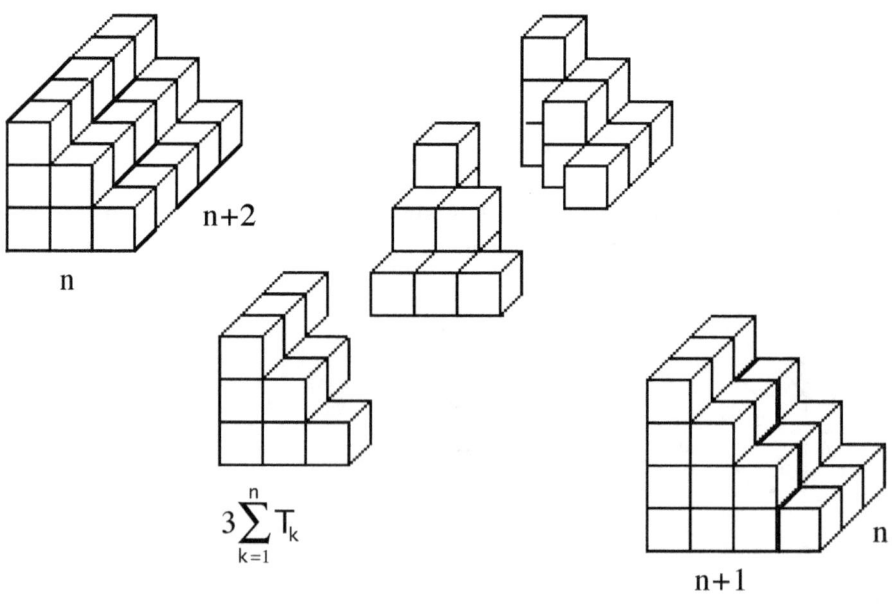

Figure 14

Suppose you have a construction like the one shown above for the case $n$. Show how it can be extended to the case $n + 1$, both in the form of the stair representing $(n+2)T_n$ and the stair representing $nT_{n+1}$.

REFERENCES

Baron, M. E. (1969). *The origins of the infinitesimal calculus*. Oxford: Pergamon.

Cupillari, A. (1989). Proof without words: $1^3 + 2^3 + 3^3 + \ldots + n^3 = [n(n+1)]^2/4$. *Mathematics Magazine, 62*, 259.

Hanna, G. (1990). Some pedagogical aspects of proof. *Interchange, 21*(1), 6 - 13.

Nelsen, R. B. (1990). Proof without words corollary: Sums of squares. *Mathematics Magazine, 63*, 314 - 315.

Wells, D. (1991). *The Penguin dictionary of curious and interesting geometry*. Penguin.

Wu, T. C. (1989). Proof without words: $(1\times2) + (2\times3) + (3\times4) + \ldots + (n-1)n = (n-1)n(n+1)/3$. *Mathematics Magazine, 62*, 27.

Zerger, M. J. (1990). Proof without words: Sums of Triangular Numbers. *Mathematics Magazine, 63*, 314.

# 17 MATHEMATICAL CONNECTIONS WITH A SPIROGRAPH[17]

A teacher can use a spirograph, a device used for artwork and designs, in the mathematics class to give students the opportunity to approach mathematics as a whole, that is, to connect ideas among different mathematical topics and with other content areas. Mathematical connections are one of the central aspects stressed in the *Curriculum and evaluation standards for school mathematics* (National Council of Teachers of Mathematics, 1989). Working with the spirograph can help students achieve several of the objectives listed in the Standard "Mathematical Connections" for grades K-12.
Students can use the spirograph to
- relate various equivalent representations of concepts to one another
- recognize, use and value relationships among different topics in mathematics;
- use a mathematical idea to further their understanding of other mathematical ideas;
- use mathematics in other curriculum areas such as art and engineering;
- explore problems and describe results using graphical, numerical, physical, algebraic, and verbal representations.

With the spirograph, which is composed of two gears,
- students can create geometrical patterns such as stars or flowers, getting a very physical, sensorial, rhythmical feeling of how they are generated;

---

[17] Flores, A. (1992). Mathematical connections with a spirograph. *Mathematics Teacher, 85,* 129-137. Copyright National Council of Teachers of Mathematics. Used by permission.

- they can use the patterns to explore mirror and rotational symmetry.
- they can study the shape of the pattern and the number of the petals of the flower, and their relationship to the number of teeth in the gears to explore concepts such as divisibility, least common multiple, ratios, and modular arithmetic.
- they can study equivalent representations using different pairs of gears, compare the patterns of the spirograph to those obtained with different materials such as string and star polygons.
- they can relate the spirograph to other settings where gears are used, such as bicycles.
- they can also be introduced to the family of curves of the hypocycloid.

Gears can be a concrete model that can help students make more abstract ideas in mathematics easier to understand. Papert (1980) vividly describes the role that gears had in his childhood as he learned mathematical topics ranging from multiplication tables to equations in two variables.

### How does the spirograph work?
The spirograph has different types of gears. Commercial spirographs can be found in toy shops. We will limit the discussion to the case when two circular gears are used, one of which turns around inside another that remains fixed (see Fig. 1).

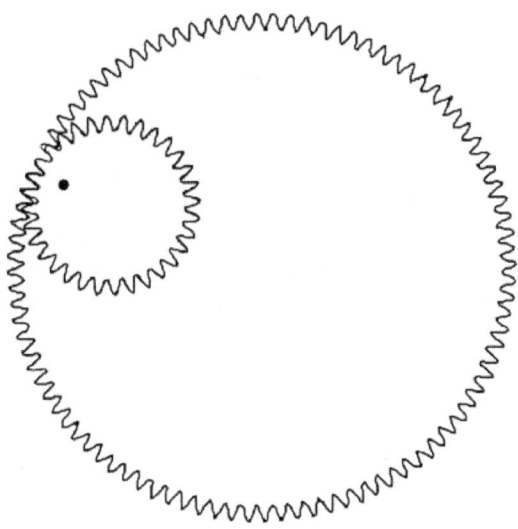

Fig. 1

As the smaller gear turns around, the spirograph traces the path of a chosen point inside the smaller gear. This way we can make geometrical patterns like stars or flowers. The number of points of the star or the flower depends on the relationship between the number of teeth of each of the gears

Let's take a given point inside the smaller gear, and start when the given point is closer to the outer gear. As the gear turns the point moves along a path, getting farther and closer to the outer gear (see Fig. 2). The point will be touching the outer gear every time the smaller gear has completed a whole turn. When the smaller gear moves all around the larger gear it is called an orbit.

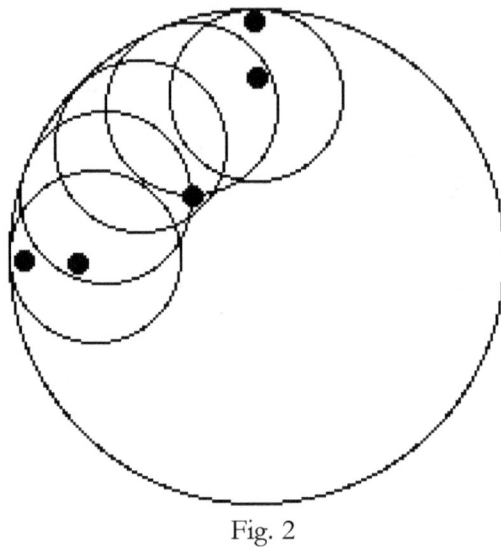

Fig. 2

If we keep turning the smaller gear inside the outer gear, a closed star (or flower) will be formed. A point of the star will be obtained for each complete turn of the smaller gear (see Fig. 3). We count one turn with respect to the center of the bigger gear, the same we count days on Earth, by measuring from the time an object faces directly to the Sun to the next time it faces to the Sun again.

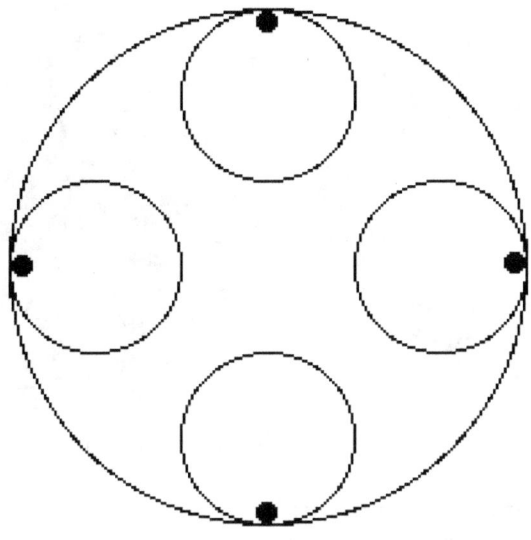

Fig. 3

## Symmetry

Explore the symmetries of the patterns in figure 4a and 4b. Describe the axis of mirror symmetry of each pattern. Describe the rotational symmetries of these patterns.

Figure 4a. Even number of petals.   Figure 4b. Odd number of petals.

We can also get patterns that have rotational symmetry without having mirror symmetry. This is done by skipping one or more gears after a star is formed, and changing the position of the tracing point in the small gear. Describe the rotational symmetries of the patterns in figure 5.

Figure 5. Patterns with rotational symmetry

**Divisibility, least common multiple.**

Let's consider the case when the outer gear has 96 teeth and the inner gear 32. When the small gear has completed a turn, it has moved along 32 teeth of the outer gear. When the smaller gear has turned three times, it has moved along 3 × 32 = 96 teeth of the outer gear, that is all the way around. The smaller gear is now again at the starting position. The path of the point is a shape with three corners (see Fig. 6). This case is very simple because 32 divides exactly into 96.

3 turns × 32 teeth = 96 teeth
1 orbit × 96 teeth = 96 teeth

Figure 6. Pattern generated with gears 96 and 32.

Let's consider another case. Take the outer gear with 96 teeth, and the smaller with 72. This time we will not return to the starting position of the smaller gear when we complete one orbit of the outer gear since the starting tooth of the small gear will not be close to the outer gear. This time we will need to orbit the small gear three times past the starting tooth of the outer gear to get a closed star with four points (Fig. 7).

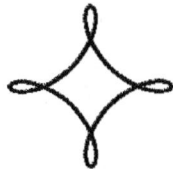

Figure 7. Pattern generated with gears 96 and 72.

When the inner gear turns 4 times the number of teeth touched will be
   4 turns × 72 teeth = 288 teeth
which is exactly the same as
   3 orbits × 96 teeth = 288 teeth

In general, with any two gears, if we keep turning the inner gear, we will eventually get a closed star, that is, we will get back to the starting position. The inner gear has orbited inside the outer gear a whole number of times, and itself has also turned a whole number of times. Thus, if we multiply the number of points of the star times the number of teeth of the small gear we get a multiple of the number of teeth of the outer gear, so it is a common multiple. It will also be the least common multiple, because when the star closes, it is the first time that we return to the starting position.

Table 1 describes this situation. The first column lists the number of teeth touched for each additional turn of the small gear. The second column lists the number or teeth touched each additional time we orbit past the starting tooth on the outer gear. The first number that appears in both columns is the least common multiple of 72 and 96. The other numbers that appear in both columns are common multiples of 72 and 96.

Table 1

|   | # teeth touched per turn | #teeth touched per orbit |
|---|---|---|
| 1 | 72 | 96 |
| 2 | 144 | 192 |
| 3 | 216 | **288** |
| 4 | **288** | 384 |
| 5 | 360 | 480 |
| 6 | 432 | **576** |
| 7 | 504 | 672 |
| 8 | **576** | 768 |

When we multiply the number of teeth of the small gear times the number of corners of the star we obtain the least common multiple of the numbers of teeth in the two gears. If the outer gear has $m$ teeth, the small gear has $n$ teeth, and the resulting star has $p$ points, the least common multiple of $m$ and $n$ will be $n \times p$.

**Example:** if $m = 92$, $n = 40$, $p = 12$ the least common multiple of 40 and 96 is $12 \times 40 = 480$.

The stars on Fig. 4a were made with the gears with the number of teeth as indicated. We can count the number of petals and complete Table 2 to obtain the least common multiple of the number of teeth in both gears.

Table 2

| $m$ | $n$ | $p$ | **least common multiple** |
|-----|-----|-----|---------------------------|
| 96  | 60  |     |                           |
| 96  | 80  |     |                           |
| 96  | 40  |     |                           |
| 96  | 64  |     |                           |
| 96  | 84  |     |                           |
| 105 | 84  |     |                           |
| 105 | 63  |     |                           |
| 105 | 60  |     |                           |
| 105 | 75  |     |                           |
| 105 | 50  |     |                           |

96, 60    96,80    96,40    96,64

96, 84    105, 84    105, 63    105, 60

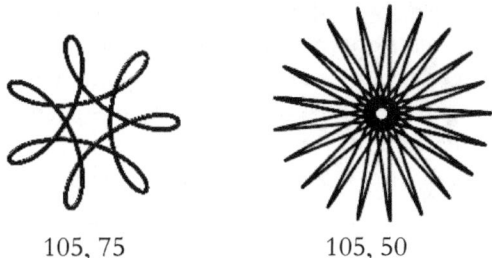

   105, 75     105, 50

Figure 4a. Number of teeth in outer gear and inner gear.

Now reverse the process. Get the least common multiple of the numbers of teeth given in Table 3 and obtain, without counting, the number of points of the stars in Fig. 4b.

Table 3

| *m* | *n* | **least common multiple** | *p* |
|---|---|---|---|
| 105 | 80 | | |
| 105 | 72 | | |
| 105 | 65 | | |
| 105 | 56 | | |

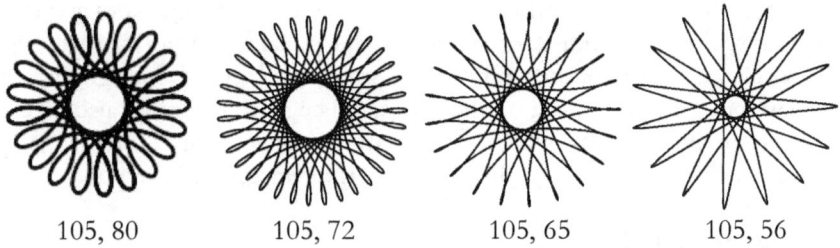

 105, 80   105, 72   105, 65   105, 56

Figure 4b. Number of petals from least common multiple

A similar relationship can be established between the least common multiple of the number of teeth in each gear, the number of teeth in the bigger gear and the number of times the smaller gear orbits past the initial tooth in the outer gear. If the outer gear has *m* teeth, the small gear has *n* teeth, and *r* the number of times we orbit past the initial tooth, the least common multiple of *m* and *n* will be *m* × *r*.

**Example:** if *m* = 96, *n* = 40, *r* = 5 the least common multiple of 40 and 96 is 5 × 96 = 480.

Students can also discover that if *k* + *n* = *m* then we get the same number of points in the star using the bigger gear with *m* teeth and a small gear with *k* teeth, as using the bigger gear with $m$ teeth and the smaller with $n$. The reason is that if $k + n = m$, then skipping $k$ teeth on the outer gear clockwise is the same as skipping $n$ teeth counter clockwise. Choosing the appropriate point inside, we can even get the same curve. This is an instance of the beautiful double generation theorem for these curves (Yates, 1974).

**Example:** if $m = 96$ and $k = 64$ we get a pattern with 3 points, the same as using $m = 96$, $n = 32$.

## Ratios

The number of petals of the flower will depend only on the ratio of the number of teeth of the two gears.

Using the gears with 105 and 70 teeth will give a pattern with 3 points, the same number of points as using the gears with 96 and 64, or any gears with ratio 3 : 2.

Students can explore what combination of gears is needed to get patterns with three, four, twelve or any desired number of points.

Spirograph patterns are closely related to star patterns (Barnett, 1981; O'Daffer & Clemens, 1976; Perl, 1981). If we have a circle with 12 points and we connect the points counting 8 points we will also get a pattern with 3 points, the same as with gears that have the same ratio.

Bicycles also use gears with different numbers of teeth. Questions such as how many wheel revolutions do we get for one pedal revolution (Ames, 1981) can be related to the number of times the smaller gear in the spirograph turns when we complete one orbit on the big gear. The spirograph can also be used to illustrate movements where combinations of rotations and orbits are present, such as planet Earth turning while it orbits around the Sun.

## Modular arithmetic

Coloring the pattern can also help to relate it to more abstract topics such as modular arithmetic, sometimes referred to as clock arithmetic (Johnson, 1961). Modular arithmetic systems are finite number systems. One way to obtain such a system is to consider the remainders of dividing different numbers by a specific fixed number. Addition of two numbers in the finite system will give a number in the system. For example, in a clock, if it is 9 o'clock and we add 8 hours, we say the result is 5 o'clock.

The pattern in figure 8 was obtained with gears with the ratio 12 : 9. We can then select a different starting point, skipping one third of the teeth between the petals, and repeat the pattern with a different color. We will get three different flowers. If we number the points (see Fig. 5) we will see that the numbers that have the same remainder when divided by 3 correspond to petals of the same color. Thus 1, 4, 7, 10 (remainder 1) correspond to a color, 2, 5, 8, 11 (remainder 2) correspond to a different one, and 0, 3, 6, 9 (remainder 0) to another one.

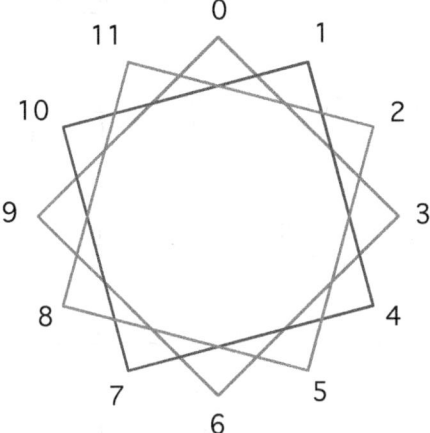

Figure 5. Remainders mod 3.

## Hypocycloid, cycloid, epicycloid.

The paths we draw with the spirograph are part of a family of curves that are generated by a point attached to a circle that rolls inside another circle. If the point is on the circumference of the inner circle, the curve is a hypocycloid. Special cases of hypocycloids are the deltoid (3 points), and the astroid (4 points). The tracing point can also be inside of the inner circle, as is the case of the spirograph. Specific ratios of the radii of the two circles give special curves such as the ellipse (ratio 2 : 1), or rose curves (ratio $2n : (n - 1)$). The hypocycloid family is related to other interesting families of curves such as the cycloid (the path of a point on a circle rolling on a straight line), and the epicyloid (the path of a point on a circle rolling outside another circle). Cycloidal curves can be obtained using the type of gears of the spirograph that have a large straight side. The properties of these curves, for example, that the cycloid is the path of quickest descent, or that it is a tautochrone, can be a topic for independent study for more advanced students (Yates, 1974).

## References

Ames, Pamela. Bring a bike to class. In <u>Activities for Junior High School and Middle School Mathematics</u>. National Council of Teachers of Mathematics, 1981. p. 78-81.

Bennet, A. B. Star patterns. In <u>Activities for Junior High School and Middle</u>

School Mathematics. National Council of Teachers of Mathematics, 1981. p. 48-50.

Johnson, D. A.; Glenn, W. H. Invitation to mathematics. Webster Pub. Co., 1961.

National Council of Teachers of Mathematics. Curriculum and evaluation standards for school mathematics. National Council of Teachers of Mathematics, 1989.

O'Daffer, P. G.; Clemens, S. R. Geometry: an investigative approach. Addison Wesley, 1976. p. 58-62.

Papert, S. Mindstorms: children, computers and powerful ideas. Basic Books, 1980.

Perl, T. String sculpture in the mathematics laboratory. In Activities for Junior High School and Middle School Mathematics. National Council of Teachers of Mathematics, 1981. p. 43-47.

Somervell, E. L. A rhythmic approach to mathematics. National Council of Teachers of Mathematics, 1975.

Yates, R. C. Curves and their properties. National Council of Teachers of Mathematics, 1974.

**Commercial spirograph**

Spirograph, design toy. Kenner Products, Cincinnati, OH. 1986.

# 18 THREE APPROACHES TO THE GOLDEN RATIO[18]

In the following examples, you will define a procedure by pushing a set of keys on the calculator, and then, by pressing ENTER again and again, the calculator will repeat the process to obtain successive terms in a sequence. That is, with the help of the calculator you will apply an iterative process to obtain the terms of the sequence. The sequence of key strokes defines an algorithm to obtain a term from the previous one, and suggests how the algorithm can be translated into a program.

a) First approach. Compute the following sequence:

1, $\sqrt{(1+1)}$, $\sqrt{(1+\sqrt{(1+1)})}$, ... That is, to obtain a new term add 1 to the previous term, and take the square root of the sum: $S_n = \sqrt{(1 + S_{n-1})}$

Key strokes for TI-83 Plus and TI-84 Plus

1   ENTER   2nd   $\sqrt{}$   1   +   2nd   ANS   )   ENTER   ENTER   ENTER ...

This sequence of keystrokes will tell the calculator to compute $\sqrt{(1 + Ans)}$, where Ans is the previous value, and every time the ENTER key is pressed

---

[18] Flores, Alfinio. Three approaches to the golden ratio. In *Calculators in Mathematics Education 1992 Yearbook* (p. 241-244.), Reston, VA: National Council of Teachers of Mathematics, 1992. Copyright National Council of Teachers of Mathematics. Used by permission.

the instruction is repeated with the new answer. The display will show

Repeat the process by pressing Enter to obtain successive terms, until the numbers on the display do not change. The result is $r = 1.618033989...$ Square this number: $r^2 = 2.618033989...$ Observe that the numbers after the decimal point are the same. Express this relationship by the equation $r^2 = 1 + r$ or by $r = \sqrt{(1 + r)}$

Notice the similarity between this last equation and the process $\sqrt{(1 + \text{Ans})}$.

b) Second approach. Let 1 be the first term of a sequence. Then, each new term is obtained by adding 1 to the reciprocal of the previous term.

$S_1 = 1$

$S_2 = 1 + 1/1$

$S_3 = 1 + \dfrac{1}{1 + 1/1}$

$S_4 = 1 + \dfrac{1}{1 + \dfrac{1}{1 + 1/1}}$

$S_5 = 1 + \dfrac{1}{1 + \dfrac{1}{1 + \dfrac{1}{1 + 1/1}}}$

The general term is: $S_n = 1 + 1/S_{n-1}$

# TO CONNECT IS TO UNDERSTAND MATHEMATICS 4

Key strokes for TI-83 or TI-84

1  ENTER  1 + 2nd  ANS  $x^{-1}$  ENTER  ENTER  ENTER ...

The calculator will display

```
1
                      1
1+Ans⁻¹
                     1.5
                       2
      1.666666667
                     1.6
              1.625
1.615384615
1.619047619
1.617647059
1.618181818
1.617977528
1.618055556
```

Repeat the process by pressing ENTER to obtain successive terms until the numbers on the display do not change.

The result is $r = 1.618033989\ldots$

Take the reciprocal of this number: $1/r = 0.618033989\ldots$

The numbers after the decimal point are the same. Express this relationship by an equation:

$r = 1 + 1/r$

Notice the similarity of this equation and the process $1 + \text{Ans}^{-1}$.

This equation is also equivalent to the equations obtained in a).

The positive root of $r^2 - r - 1 = 0$ is the limit of the two previous sequences, and is known as the golden ratio.

c) Use the MATH key in calculator to convert each term in the previous sequence to a fraction to express the terms of the sequence defined in b) as fractions.

In the previous sequence of keystrokes, you need to add a step, press MATH and press ENTER to choose the first option ▶ Frac

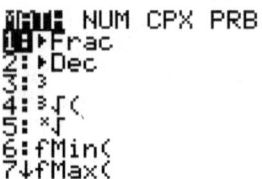

The calculator will show the corresponding fraction.

Key strokes for TI-83 or TI-84

1   ENTER   1   +   2nd   ANS   $x^{-1}$   MATH   ENTER   ENTER   ENTER...

```
1
1+Ans⁻¹▶Frac
              1
              2
            3/2
            5/3
            8/5

          13/8
          21/13
          34/21
          55/34
          89/55
         144/89
         233/144
```

The sequence of fractions is:

$$\frac{1}{1}, \frac{2}{1}, \frac{3}{2}, \frac{5}{3}, \frac{8}{5}, \frac{13}{8}, \frac{21}{13}, \frac{34}{21}, \frac{55}{34}, \ldots$$

The denominators and the numerators are the Fibonacci numbers:

1, 1, 2, 3, 5, 8, 13, 21, 34, 55, 89, 144, 233, 377, 610, . . .

Fibonacci numbers $F_n$ after the first two are obtained by adding the two previous terms $F_n = F_{n-1} + F_{n-2}$. Thus, to approximate the golden ratio, we divide a term of the Fibonacci sequence by the preceding term:

$S_n = F_n/F_{n-1}$

# 19 CALCULATORS IN CALCULUS: THAT'S THE LIMIT[19]

**Part 1 Preliminary explorations with sequences**
The following sequences can be explored by repeatedly pushing a single key after a number has been entered, or by using the number repeatedly as constant factor.

**Activity 1. Constant factor.**
Most calculators have the capability to use the same number as a constant factor (or addend or whatever). Some have a special key for that, others do not need one. The following examples will be illustrated for a calculator that remembers the last set of instructions (TI-84). They are given to convey the iterative process to calculate the sequences. The keys to be pushed may vary with other types of calculators.

a) Push the following keys in your calculator.
Key strokes for TI-83 and TI-84
2 ENTER × 2 ENTER ENTER ENTER ...

```
2
Ans*2
            2
            4
            8
           16
           32
```

Write the number on the display after pushing each set of keys:

---

[19] Flores, A. (1991). Calculators in calculus: that's the limit. *PRIMUS*, *1*(3), 295 - 301. Reprinted by permission of Taylor & Francis (http://www.tandfonline.com).

$S_1 = 2$      $S_9 = 512$
$S_2 = 4$      $S_{10} = 1024$
$S_3 = 8$      $S_{11} = 2048$
$S_4 = 16$     $S_{12} = 4096$
$S_5 = 32$     $S_{13} = 8192$
$S_6 = 64$     $S_{14} = 16384$
$S_7 = 128$    $S_{15} = 32768$
$S_8 = 256$    $S_{16} = 65536$
. . .

These numbers appear in the famous problem of the chessboard, or in the number of sheets when you fold a paper in half repeatedly.

b) Start with a different number. Write down your results.

What happens if you start with 1? _____ Why?

What happens if you start with a number between 0 and 1? _____

Try 0.1
. 1   ENTER  ×  . 1   ENTER   ENTER   ENTER ...

```
.1
Ans*.1
              .1
              .01
              .001
              1E-4
              1E-5
■
```

Write down the number on the display after each execution
0.1
0.01
0.001
0.0001
. . .

To what number do these numbers approach? _____. Notice that we can come as close to 0 as we want. 0 is the limit of this sequence.

Use other numbers to start with. Try 0.9
Key strokes for TI-84
. 9   ENTER  ×  . 9   ENTER   ENTER   ENTER ...

# TO CONNECT IS TO UNDERSTAND MATHEMATICS 4

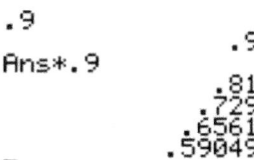

■

Write down the number on the display after each execution
0.9
0.81
0.729
0.6561
0.59049
. . .

To what number do these numbers approach? _____

What keys do you have to push to obtain the following sequences?

0.99, 0.992, 0.993, 0.994, . . .

1.00001, 1.000012, 1.000013, 1.000014, . . .

What happens if you start with 0? _____

What happens if you start with a negative number? Try first a number between -1 and 0. Try a number less than -1.

Represent the sequence by algebraic expressions, assuming the first term is $a$.
What can you say about the sequence
$a, a^2, a^3, a^4, \ldots$    $(0 \leq a < 1)$?

What can you say about the sequence
$a, a^2, a^3, a^4, \ldots$    $(1 < a)$?

## Activity 2
Push the following keys:

2 ENTER $x^2$ ENTER ENTER ENTER ENTER . . .

What numbers appear on the display? Write them down.
2, 4, 16, 256, 65536, . . .

What is the relationship between these numbers and those in the first example of Activity 1a)?

How quickly do you reach numbers bigger than the calculator is able to deal with? _____
Each number is the square of the previous one. Represent this sequence using exponents. What is the relationship of the exponents and the numbers in the first example of Activity 1a)?

Start with a different number. Choose a number that is very close to 1, for example 1.000001
1.000001 ENTER $x^2$ ENTER ENTER ENTER ENTER …

What are the numbers of the sequence this time?

1.000001
1.000002
1.000004
1.000008
1.000016
1.000032
. . .

Now choose a number close to 1 but smaller, for example, 0.999999
What happens with the numbers of the sequence in this case?

.999999 ENTER $x^2$ ENTER ENTER ENTER ENTER …

0.999999
0.999998
0.999996
0.999992
0.999984

Represent this type of sequence with an algebraic expression. Use exponents
The sequence can also be defined by

$S_n = S_{n-1} * S_{n-1}$        $S_1 = 0.99999$

or in general

$S_n = S_{n-1} * S_{n-1}$        $S_1 = a \ (0 \leq a < 1)$

What relationship is there between this sequence and the one on Activity 1? _____ How much faster does this subsequence converge?

## Activity 3
Push the following keys:
2 ENTER 2ND √ 2ND ANS ) ENTER ENTER ENTER ENTER ...

Observe the numbers that appear on the display after each execution
2
1.4142...
1.18...
1.09...
1.04...
1.02...
To what number do they approach?
The sequence you are exploring is

2, $\sqrt{2}$, $\sqrt{\sqrt{2}}$, $\sqrt{\sqrt{\sqrt{2}}}$, ...

Express this sequence using fractional exponents
(remember that $\sqrt{x} = x^{1/2}$)

What happens if you start with a much bigger number, for example 1000000?
1000000 ENTER 2ND √ 2ND ANS ) ENTER ENTER ENTER ...

Write down the numbers.
1000000
1000
31.62...
5.62...
2.37...
1.53...
Do they approach a number? _____

What happens if you start with 1? _____
Or with 0? _____

What happens if you start with a number between 0 and 1, for example 0.000001?

0.000001 ENTER 2ND √ 2ND ANS ) ENTER ENTER ENTER ENTER ...

0.000001
0.001
0.0316...
0.177...
0.42...
0.64...
0.80...
0.89...

Write the terms of this sequence in algebraic notation.

# 20 A PUZZLE OF MATHEMATICAL FORMULAS[20]

ABSTRACT: 24 of the most beautiful and important formulas in undergraduate mathematics are assembled to form a puzzle.

KEYWORDS: Mathematical formulas, puzzles

INSTRUCTIONS
Photocopy the puzzle in the following two pages. Cut along the lines, shuffle the pieces and reassemble. When the puzzle is formed, equivalent expressions will be opposite to each other along the same edge. If we break the puzzle vertically down the middle, the right-hand side of the second page should match with the left-hand side of the third page.

Most formulas should be familiar to college students. Although my students did not know all the formulas, they worked around the ones they knew, and were able to form it.

Some formulas need explanation, like . The figures represent a cone, a semi-sphere, and a cylinder with the same base and the same height, their volumes are in proportion 1 : 2 : 3. In Euler's formula $V + F = E + 2$, V is the number of vertices, F of faces, E of edges of a convex polyhedron. Some of the formulas have a deeper role in mathematics than would at first strike the eye, for example, Halmos calls the geometric series $1/(1-x) = 1 + x + x^2 + x^3 + ...$
one of the (computational) elements of mathematics.

---

[20] Flores, A. (1991). A puzzle of mathematical formulas. *PRIMUS, 1*(4), 397 - 400. Reprinted by permission of Taylor & Francis (http://www.tandfonline.com).

Students will find old friends, like log $(ab)$ = log $a$ + log $b$ (do they know that log is an isomorphism of groups?) Some formulas establish surprising connections between the important mathematical constants $\pi$, $i$, $e$, 1, 0. Somebody said that the formula $e^{i\pi} + 1 = 0$ is a proof that God is a mathematician. Of course, this formula is just a particular case of the formula that connects the exponential, sine and cosine functions. Some formulas may not be known to your students. While forming the puzzle, they will see that there are sides that will have to match, and they do not know what the material is. They will ask questions. Use this opportunity to point to them fields of mathematics not yet known to them, and where to find more about them. Most formulas in this puzzle are associated to famous mathematicians: Gauss, Euler, Archimedes, Pythagoras... You can modify this puzzle, other edges can be well known integrals, derivative formulas, functions and their curves, characterizations of spaces, or of topologies.

REFERENCE
Halmos, P. R. 1981. Does mathematics have elements? *Mathematical Intelligencer*, 3(4): 147- 153.

BIOGRAPHICAL SKETCH
Alfinio Flores studied mathematics in Mexico City (spicy food and people, earthquakes, crowds, traffic jams, and beautiful weather). He studied mathematics education at Ohio State University. His background has prepared him to live in California. He uses computers, calculators, concrete materials, and any other means to teach mathematics at San Diego State University.

# TO CONNECT IS TO UNDERSTAND MATHEMATICS 4

| | |
|---|---|
| [small square figure] | [rotated square figure]    not $(P \Rightarrow Q)$ |
| $1 + x + x^2 + \ldots + x^n$ | $1 + 3 + 5 + \ldots + (2n-1)$ |
| $\dfrac{1 - x^{n+1}}{1 - x}$    $\dfrac{\pi^2}{6}$    $\displaystyle\sum_{i=0}^{n} \binom{n}{i} a^{n-i} b^i$ | $\dfrac{1}{1^2} + \dfrac{1}{2^2} + \dfrac{1}{3^2} + \dfrac{1}{4^2} + \ldots$    $n^2$    $e^{i\pi}$    $V + F$ |
| $(a + b)^n$    $\dfrac{d(4\pi r^3/3)}{dr}$    $\dfrac{d\,e^x}{dx}$ | $E + 2$    $4\pi r^2$    $e^{ix}$    $\sin^2 + \cos^2$ |
| $e^x$    $1^3 + 2^3 + 3^3 + \ldots + n^3$ | $\cos x + i \sin x$    $\left[\dfrac{n(n+1)}{2}\right]^2$    $\log(ab)$ |

| | |
|---|---|
| $\int_1^e \frac{1}{x}\,dx$ <br><br> $\frac{\pi}{4}$ | P and not Q <br><br> 1 <br><br> $1 + 2 + 3 + \ldots + n$ |
| $\frac{1}{1} - \frac{1}{3} + \frac{1}{5} - \frac{1}{7} + \frac{1}{9} - \ldots$ <br><br> 0 <br><br> $\binom{n}{r+1} + \binom{n}{r}$ | $\frac{n(n+1)}{2}$ <br><br> $-1$ <br><br> $\log 1$ <br><br> not P  or  not Q |
| $\binom{n+1}{r+1}$ <br><br> $\frac{d(\pi r^2)}{dr}$ <br><br> $(\cos\theta + i\sin\theta)^n$ | not ( P and Q ) <br><br> 1 <br><br> $2\pi r$ <br><br> △ ⌒ ▱ |
| $\cos n\theta + i \sin n\theta$ <br><br> $-1$ | $1 : 2 : 3$ <br><br> $i^2$ <br><br> $\log a + \log b$ |

# 21 ESTIMATION PERFORMANCE AND STRATEGY USE OF MEXICAN 5TH AND 8TH GRADE SAMPLE[21]

ABSTRACT. What computational estimation skills and strategies do Mexican students possess? Does the theoretical model based on interviews with a select United States sample accurately describe the Mexican sample? These were questions studied based on interviews with 8 eighth graders (those scoring in the top 5%) out of a sample of 177 eighth graders from twelve different Mexican schools representing a range of social and economic backgrounds. Preliminary screening data collected by administering a computational estimation test revealed that estimation was very difficult for the Mexican students (mean 4.0, range of 0 to 18 on the 38-item open-ended test).
The interviews revealed that the Mexican students as a whole did employ the three general cognitive processes outlined in the theoretical model, namely reformulation, translation, and compensation. The most common strategy employed was the front-end technique. Similarly, a frequent strategy used to "estimate" was mentally applying a paper/pencil algorithm. In contrast to data collected under similar conditions in Japan and the United States, rounding was a strategy only occasionally used in the interviews. The use of benchmarks (key reference points used as bounds in forming an estimate) as a strategy for estimating problems involving percent was common and may reflect students' "out-of-school" experience with mathematical applications. Consistent with parallel investigations with Japanese and United States

---

[21] Reys, B. J., Reys, R. E., and Flores Peñafiel, A. (1991). Estimation performance and strategy use of Mexican 5th and 8th grade sample. *Educational Studies in Mathematics, 22*, 353-375. Used by kind permission of Springer.

students, these Mexican students rarely reflected on their estimates through their own initiative and rarely recognized unreasonable estimates.

INTRODUCTION

*Background on Estimation*
Computational estimation has been recognized as an important mathematics topic, emphasized by professional organizations, and identified in recent curricular recommendations in several countries, including the United States, England, and Japan (Conference Board of the Mathematical Sciences, 1989; National Council of Teachers of Mathematics, 1989; National Council of Supervisors of Mathematics, 1989; Cockroft, 1982; Japan Ministry of Education, 1989). However, computational estimation has not been taught or investigated in Mexican schools. Previous research in the United States (Reys et al., 1982) and in Japan (Reys et al., 1991) has provided a framework about how good estimators produce estimates. Three general cognitive processes were identified among good estimators: reformulation (changing the numerical data into a more manageable form), translation (changing the mathematical structure of the problem to a more mentally manageable form), and compensation (adjusting the estimate to correct changes due to reformulation or translation). Several specific thinking strategies for each process were identified. Here are some examples.

Reformulation
*Traditional rounding*: round each addend to the nearest multiple of a specific power of ten, and add rounded numbers.
*Front-end technique*: compute using the lead digits and determine ap-propriate place value.
*Substitution*: use equivalent form (fraction, decimal, percent).

Translation
*Averaging*: estimate an average and use it to translate from an addition problem to a multiplication problem.
*Benchmarks*: use key reference points as bounds in forming an estimate.

Compensation
*Final compensation*: adjusting an initial estimate to more closely convey the user's knowledge of the error introduced by the strategy employed.
*Intermediate compensation*: adjusting numerical values prior to them being operated on to systematically correct for error.

Research in the United States and Japan has reported that these strategies are seldom taught in school. In fact, good estimators self-develop most of the strategies they use.

Background on Mexican Education

One of the purposes of this study was to replicate previous research conducted in the United States and Japan in a different cultural and educational setting. Mexico is a country satisfying these criteria. First, it is a developing country, not a highly industrialized country like the United States and Japan. Second, Mexican culture is an amalgam of Indian and Spanish cultures, very different from the cultures of both Japan and the United States. Third, it is characterized by sharp contrasts and inequalities: a country with ancient traditions, yet one in which the majority of the population is under 18 years of age; a country whose population has doubled twice in less than 50 years, making it the second largest country in Latin America and the largest Spanish speaking country in the world.

Mexico's educational system has expanded with the population (Solana et al., 1981). Compulsory education begins with first grade at age six and continues through grade six. Education is free to students in grades K-9. However, the drop-out rate is high. For example, in the state of Guanajuato only about 65% of students finish elementary school (grade 6) and about 30% complete 9th grade.

Mexico's educational system is highly centralized. Decisions are made centrally and implemented through a hierarchical structure. The mathematics curriculum in elementary school (grades 1-6) is specified by the Secretaría de Educación Pública (the national Ministry of Education), and a single government-produced textbook for each grade is used throughout the country. In grades 7-9, textbooks are published by private companies, but must follow the national curriculum and be approved by the Ministry of Education. Despite the centralized structure there is still wide variability in schools and classrooms. It is not possible to describe the variety of Mexican schools in a few statements; however, the following observable characteristics are common and would immediately strike a foreign visitor:
- teachers lecture for most of the class time
- classes are frequently large (40-50)
- many schools have two shifts
- many teachers work two shifts or at several schools
- supplementary materials, including worksheets are not used
- technology is not used (except for television in special schools)

With the growing population, it has been easier to satisfy the demand for school buildings than for qualified teachers, particularly in mathematics. In an effort to provide uniform instruction on core curricular topics, the state system has implemented the use of television in grades 7-9 in about 400 schools in the state of Guanajuato. In these schools a single teacher teaches all subjects, including mathematics, science, social science, Spanish and English. Seventeen minute lessons in each subject area are broadcast nationally from Mexico City. Following the broadcast students ask questions and do further work with the help of the teacher and textbook. About 85% of the teachers in TV schools in Guanajuato do not have any special preparation as mathematics teachers.

Purpose of the Study
The primary purpose of this study was to identify and characterize the computational estimation skills and strategies of Mexican students. A second purpose was to add knowledge to the development of a general framework that characterizes the thinking processes used by successful estimators. Specifically, the computational estimation processes used by the best estimators in grades 5 and 8 were explored. The framework constructed from interview data with a sample of Mexican students was compared with earlier frameworks proposed to describe processes used by good estimators in the Unites States and Japan.

METHOD

Procedures

The basic research design was modeled after the plan used in earlier studies (Reys et al., 1982; Reys et al., 1991). A computational estimation test was used to screen 432 Mexican students in order to identify students with the best computational estimation skills. As in the earlier studies, students scoring in the top 5% of 8th grade level (8 students) were defined to be 'good estimators" and scheduled for a follow-up interview. The same criterion was to be used with 5th graders, but their performance on the computational estimation test was very low (high score: 7 out of 38 items), compared to that of the eighth-grade sample. Two interviews conducted with the top-scoring fifth graders confirmed that the estimation tasks were very difficult. Consequently, no further interviews were conducted with this group. All interviews were conducted during the regular school day within three weeks of the administration of the screening test.

Subjects
Twelve different Mexican schools (6 elementary, 6 secondary) participated in this research. All schools were within the State of Guanajuato, located in the interior of Mexico. School districts were selected to represent a range of social and economic backgrounds, and to be as representative of the overall population as possible. One-third of the schools were within the city of Guanajuato, which is the state capital and also home of the State University,

and the Teacher's College. Another one-third of the schools were selected from Leon, the largest metropolitan area (population over 1,000,000), and the major industrial center in the state of Guanajuato. The remaining one-third of the schools were rural and situated in or near Silao. All schools were within 50 km of the city of Guanajuato. Half of the secondary schools selected were television schools, and the other half were regular schools. Since most of the schools in Mexico run two shifts in order to accommodate all of the students, half of the schools were selected from each shift. All schools were randomly selected within their district. Once a school was selected, one class within the school was randomly selected to participate. The class size ranged from 31 to 54 students in grade five and from 20 to 43 in grade eight. Students in all classes were heterogeneously grouped as is the custom in Mexican schools. However, the age of students in the grades ranged from 10 to 12 for the fifth-grade classes, and from 12 to 17 for the eighth-grade classes. The sampling resulted in a pool of 255 fifth graders and 177 eighth graders.

Instruments

*Estimation test.* The 38 open-ended-item estimation test used in this research contained 25 items from the Assessing Computational Estimation test (Reys et al., 1980). Thirteen items from the Assessing Computational Estimation test were modified slightly to make them appropriate for Mexican students (e.g. prices changed from dollars to pesos to reflect realistic values). One new item (Estimate: $12/13 + 7/8$) from the Second National Assessment of Educational Progress was added because it is an item on which some national data are available in the United States (Carpenter et al., 1980).
Each of the test items was produced on a 35-mm slide and shown using a carousel slide projector. This format allowed for group administration and controlled the amount of response time (12-15 seconds) for each item. The test included 25 straight computation items (containing only numerical data) and 13 application items (containing numerical data embedded in a physical context) designed to be relevant to Mexican students, and pre-sented in Spanish. All four operations were included but the majority of items involved multiplication and division. A few items included fractions and decimals, but most used whole numbers.
Open-ended responses have been shown to provide a more valid measure of computational estimation (Reys et al., 1980) because they provide no clues about the answer, and because their nature guarantees that a range of estimates will be produced. In order to score the open-ended questions, the researchers established "acceptable intervals" for each of the 38 items. This was done by identifying strategies appropriate for answering the questions and then determining the upper and lower values of their resulting re-sponses. The least-lower, and the greatest-upper values were used as

endpoints for the acceptable intervals. The procedure used to establish the intervals for acceptable estimates reflects research-based recommendations (Reys, 1986).

At the conclusion of the estimation test the students were asked two attitudinal questions which focused on their self-perception of estimation:

Are you a good estimator?
Do you think estimation is important?

In each case, students were to choose among a choice of three responses (yes, no, not sure), the response that best reflected their feelings.

Uniform instructions for the estimation test were used in each school. Test administrators were Mexican teachers with wide experience with students of those ages, and who were taking part in the graduate program in Mathematics Education at the University of Guanajuato. Students were told that this was an estimation test. The meaning of the word estimation was carefully explained, using several equivalent words or phrases, such as "make a rough calculation, get an approximate answer" [resultado aproximado, no la respuesta exacta, sino más o menos]. Students were also told that each problem would be timed, and that they would have between 12 and 15 seconds to make and record their estimate. Time was restricted because prior research has demonstrated that valid measures of computational estimation in a group setting are very difficult to obtain without controlling the time (Reys, 1988). The students were also told "do not copy the problem, do the work in your head." Prior to starting the test, two sample problems were provided to acquaint students with the format and the time restriction of each item. Immediately following these two sample items, the students were shown acceptable intervals for each item and told that any estimate within this interval would be considered a good estimate. There was no teaching of appropriate estimation techniques, and no discussion of the criteria used to establish the acceptable intervals. These two sample items were designed to orient students to the estimation test, and to allow students to adjust their seats to see the screen clearly and ask for further clarification of the task, if necessary.

*Interview protocols.* Ten items (four straight computation, and six applied computation) were formulated for the interview. Five of these items appeared on the estimation test (see Table I). These items were reviewed by the research team for consistency with the Mexican national mathematics curriculum and judged appropriate for both the fifth and eighth grade levels. In order to gain as much insight into the students' thinking processes as possible, no time restrictions were placed on the interview items. Students were allowed as much time as needed to formulate their estimates and describe how they were produced.

Table I
Estimation interview items
  Straight computation:
  1.  87,419
      92,765
      90,045
      81,974
    + 98,102

  2.  8,127 $\overline{)474,257}$
  3.  $\dfrac{12}{13} + \dfrac{7}{8}$
  4.  4.486 × 0.24

  Applied computation:
  5.  How much area does this rectangle contain? 28 × 47
  6.  If 30% of the fans at the game buy a soda, about how many sodas are sold? Attendance: 54,215
  7.  During the Mexico World Series, concessions netted $21,319,908. If these proceeds are to be equally divided among the 26 teams, how much does each team receive?
  8.  This grocery store ticket has not been totalled. Estimate the total. _____

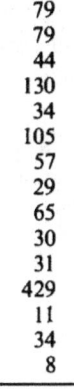

  9.  Estimate the discount on the jacket.

      $28,000
      15% discount

  10. The student missed two of these problems. Which are unreasonable?

      1. $\dfrac{4}{9} + \dfrac{5}{10} = 1\dfrac{5}{90}$

      2. $\dfrac{5}{8} + \dfrac{4}{7} = \dfrac{8}{15}$

      3. $\dfrac{8}{15} + \dfrac{11}{20} = 1\dfrac{1}{12}$

To ensure consistency in the presentation of the interviews, specific probes were developed. The interview items and probes were assembled into an interview packet. Training sessions using videotaped student interviews as well as discussions of interview techniques were held. The interviewers also conducted several pilot interviews prior to conducting the planned interviews. Discussion and clarification followed these pilot interviews. The interviews were conducted by two interview teams, each team consisting of three members - an interviewer, an observer, and a camera person.

Individual interviews were conducted with 10 of the best estimators (two 5th and eight 8th graders) to ascertain the strategies and processes used while estimating. They were selected based on performance on the estimation test. Each of these students scored in the top 5 percent in their respective grades on the estimation test. Students were asked to describe the strategies and thought processes they used while deriving estimates to the ten individually presented items. Following the completion of two trial fifth grade interviews and a review of the videotape and estimation test data, a decision was made to discontinue further interviews with fifth graders. These students had difficulty with each of the interview items and generally lacked conceptual knowledge sufficient to establish estimation techniques. Consequently, the interview data reported here consists only of the eight 8th grade interviews.

The length of the interviews ranged from 35 to 80 minutes. All interviews were videotape-recorded, and a transcript was made of the student comments judged particularly interesting and insightful. Upon completion of an interview the researchers met with the interview team, reviewed the videotape, discussed and categorized techniques and strategies used on each item and completed a summary interview form.

RESULTS

Estimation test

Table II summarizes the total scores on the 38-item estimation test for all fifth and eighth grade students. In grade five, the scores ranged from 0 to 7, with a mean of 1.1, and standard deviation of 1.5. In grade eight, the scores were more widely distributed, ranging from 0 to 18, with a mean of 4.0, and standard deviation of 3.2. As expected, there was a significant ($p < .01$) grade level difference. There was however not any significant difference ($p > .4$) in performance between males and females in either grade. This contrasts with consistent gender differences on estimation performance across all age groups in research conducted with a United States sample (Reys et al., 1982; Rubenstein, 1985).

Table II
Summary of scores on 38-item computational estimation test

| Score | Grade 5 N = 255 Frequency | Percent | Grade 8 N = 177 Frequency | Percent |
|---|---|---|---|---|
| 0 | 121 | 47.5 | 14 | 7.9 |
| 1 | 58 | 22.7 | 26 | 14.7 |
| 2 | 39 | 15.3 | 35 | 19.7 |
| 3 | 12 | 4.7 | 19 | 10.7 |
| 4 | 10 | 3.9 | 23 | 13.0 |
| 5 | 12 | 4.7 | 14 | 7.9 |
| 6 | 2 | 0.8 | 9 | 5.1 |
| 7 | 1 | 0.4 | 15 | 8.5 |
| 8 | 0 | 0 | 6 | 3.4 |
| 9 | 0 | 0 | 8 | 4.5 |
| 10 | 0 | 0 | 1 | 0.6 |
| 11 | 0 | 0 | 0 | 0 |
| 12 | 0 | 0 | 3 | 1.7 |
| 13 | 0 | 0 | 2 | 1.1 |
| 14 | 0 | 0 | 0 | 0 |
| 15 | 0 | 0 | 0 | 0 |
| 16 | 0 | 0 | 1 | 0.6 |
| 17 | 0 | 0 | 0 | 0 |
| 18 | 0 | 0 | 1 | 0.6 |

The estimation screening test was the tool used to identify the best estimators within the sample. It should be understood that the use of the term "good" to describe "good estimators" is relative to this group of Mexican students. Even though the focus of this article is on the interview data, the test results provide an important source of information. There were many similar patterns of performance between the fifth and eighth grade students, including a correlation of 0.87 between the performance on items for grades five and eight. Students in both grade levels did better on the application problems than on the straight computation items.

Performance levels on five of the easiest and five of the most difficult items are provided in Tables III and IV. Several different operations with whole numbers, fractions and decimals are also included in Tables III and IV to provide an overall perspective of student performance. The items shown in Table III are those answered correctly by more than 10 percent of the fifth graders and by more than 25 percent of the eighth graders. Although performance among eighth graders was consistently higher on all items, the performance of students in both grades was very low.

Table III
Five of the easiest estimation test items

| Item | | Acceptable interval |
|---|---|---|
| A. | About how long are these altogether?  [ 39 ] [ 99 ] [ 37 ] | 160–200 |
| B. | 8)713 | 70–100 |
| C. | 4 × 7.47 | 28–32 |
| D. | 61.3 × 0.8 | 48–61.3 |
| E. | About what is the difference in populations?  San Martin 12,367  Silio 3,788 | 8000–9000 |

Percent correct by fifth and eighth grade students

| Item | Grade 5 | Grade 8 |
|---|---|---|
| A | 10.2 | 37.8 |
| B | 20.0 | 37.3 |
| C | 12.6 | 30.0 |
| D | 4.7 | 22.0 |
| E | 2.4 | 16.4 |

Among the attitudinal questions on the screening test were the questions, "Are you a good estimator?" and "Do you think estimation is important?" A summary of responses is shown in Table V. A high percent of students at each grade level said estimation is important, but only a few students rated themselves as good estimators.

The best estimators were defined to be students scoring in the top 5% on the estimation test. The low and small range of scores in both grades reveals that the test was very difficult. Table II confirms that the screening test produced a wider range of scores for the eighth-grade sample (0 to 18), but after the two high scores of 18 and 16 there was a drop to 13. The fifth-grade high score was 7, and about 70% of the fifth graders scored zero or 1.

TABLE IV
Five of the most difficult estimation test items

| Item | | Acceptable interval |
|---|---|---|
| F. | 87,419<br>92,765<br>90,045<br>81,974<br>+ 98,102 | 430,000–460,000 |
| G. | $0.7 + 0.002 + 0.81$ | 1–2 |
| H. | $98.6 \times 0.041$ | 3.5–4.1 |
| I. | $12/13 + 7/8$ | 1.5–2 |
| J. | A motorbike travels 1322 km on 48 l of gas. About how many km per liter of gas? | 20–30 |

Percent correct by fifth and eighth grade students

| Item | Grade 5 | Grade 8 |
|---|---|---|
| F | 0 | 0 |
| G | 0 | 2.2 |
| H | 0.8 | 0.6 |
| I | 0.8 | 3.4 |
| J | 1.6 | 3.4 |

Interview

General observations. Following the interviews, the summary interview forms for the eighth graders were studied and a list of strategies for each of the ten items compiled. Using the individual student summary interview forms and the summary-by-item data, the researchers compiled a list of general observations:

TABLE V
Percent of students' responses to attitudinal questions related to estimation

| Statement | Grade | Yes | No | Not sure | No response |
|---|---|---|---|---|---|
| Are you a good estimator? | 5th<br>8th | 31.0<br>15.2 | 20.5<br>28.8 | 28.0<br>53.7 | 20.5<br>2.3 |
| Do you think estimation is important? | 5th<br>8th | 70.6<br>91.0 | 10.2<br>4.5 | 4.7<br>2.3 | 14.5<br>2.2 |

- The front-end estimation strategy was the most common technique

observed. This strategy was used by eight students and was most often applied by six of the eight.
• Traditional rounding as a strategy for estimating was used by three of the eight students interviewed and only one of these used this strategy consistently
• The idea of compatible numbers was not verbalized, (i.e., was not described by students as a factor they considered when reformulating numbers). However, in at least four interviews, students chose rounded numbers which result in compatible sets of numbers.
• The use of benchmarks was apparent in six interviews, particularly for the items dealing with percent.
• There was a tendency on the part of at least five students to use mental computation to produce an exact answer rather than an estimate. Fre-quent errors (both procedural and computational) occurred during the use of mental computation.
• Misconceptions surrounding decimals and fractions were evident as five of the eight students formed estimates.
• Conceptual understanding of percent was apparent in all students, and led to efficient use of benchmarks when estimating with percents.
• Only four students answered affirmatively to the question "can you think of another way to make an estimate for this problem?"
• Reformulation was used by all students, translation by six, and compen-sation by 7 students.
• Seven of the students never reflected on their estimates once they were produced, and did not notice unreasonable estimates.
• From the students' reports, estimation is not taught, learned or used in school. A few students did indicate they developed some estimation skills in out-of-school consumer-type settings (for example, percent discounts used in stores).

Each of these general observations will be briefly discussed with examples from student interviews used to clarify the observation. Due to space considerations, only selected interview items are summarized and discussed here, namely those which describe and highlight some of the general processes observed in the interviews.

*Strategy use.* The interviews were characterized by a limited variety of estimation strategies. Among the reformulation strategies, front-end, rounding, and compatible numbers were described by students, although they did not have any concise vocabulary to discuss the techniques being used. The most common strategy employed was the front-end technique, used by all students. For example, on item # 1, addition of five numbers between 81,000 and 99,000 (see Table I), one student described his strategy in this way:

"Add the first column - that is 9 × 3 + 8 + 8 or 43 so it is 430,000 but really a little more. Let's say 450,000." There appeared to be no specific procedure used to make the final adjustment, rather it reflected the stu-dent's "gut level" feeling for the error surrounding his estimate.

The front-end strategy was also the most popular strategy for estimating the total of the grocery store ticket (item # 8 in Table I). Although it was the most popular strategy employed for these two multi-digit addition items, it did not result in a high success rate. For example, in item # 1, only 1 of the 8 students produced an estimate within the acceptable range. The strategies used and the resulting estimates for item # 1 are summa-rized in Table VI. Order-of-magnitude errors accounted for 4 of the 7 errors while the other 3 responses were just outside the acceptable interval.

Rounding was a strategy only occasionally used in the interviews. On item # I none of the 8 students used a traditional rounding strategy (round each addend to the nearest hundred-thousand or ten-thousand). Interview item # 5 (estimate the area of a 28 by 47 rectangle) might be considered the most likely candidate for use of a traditional rounding strategy. However, only 3 students employed this strategy (all successfully), while 4 students employed mental computation using the exact data to produce their "estimate" (3 unsuccessfully, 1 successfully). One student misinterpreted area as perimeter. In general, the lack of use of the rounding strategy during these interviews is in marked contrast to similar data collected in both the United States and Japan where students made extensive use of this school-taught technique. Rounding, however is not taught as an estimation skill in Mexican schools.

TABLE VI
Summary of interview data for item # I

| Strategy | Frequency | Estimate(s) | Time (sec.) |
|---|---|---|---|
| Front-end with adjustment | 5 | 500,000 | 25 |
| | | 450,000* | 76 |
| | | 42,000 | 30 |
| | | 475 | 26 |
| | | 500,000 | 24 |
| Averaging | 3 | 45,000 | 109 |
| | | 550,000 | 16 |
| | | 45,000 | 27 |

* Denotes estimate within the acceptable interval established by the researchers for this estimation item.

## TABLE VII
Examples of use of benchmarks as a strategy for estimating

| Interview item | Estimate | Explanation of strategy |
|---|---|---|
| 7. 21,319,908 ÷ 26 | 750,000 | If there were $26 million then each team would get $1 million but it's less so they get less than $1 million, about 3/4 of a million. |
| 9. 15% of 28,000 | 5000 | 100% is 28,000; 50% is half or 14,000 25% is half again or 7000. 15% is less so adjusts to 5000. |
| 10. Which are unreasonable? 1. $\frac{4}{9}+\frac{5}{10}=1\frac{5}{90}$ 2. $\frac{5}{8}+\frac{4}{7}=\frac{8}{15}$ 3. $\frac{8}{15}+\frac{11}{20}=1\frac{1}{12}$ | | The first answer is unreasonable because adding less than 1/2 to 1/2 should give something less than 1. |

The use of benchmarks as a strategy for estimating is illustrated by examples presented in Table VII. Five of the eight students interviewed used the idea of benchmarks on item # 7, comparing the divisor (26) to the front digits of the dividend (21) and remarking that it "must be less than one million." All five students went further by stating an adjusted estimate less than one million. The use of benchmarks when estimating with percents will be discussed in the next section on conceptual understanding.

One of the most common strategies used to produce an "estimate" was mentally applying a paper/pencil algorithm. This process seemed to be the first choice of some students. In fact, the highest scoring student on the estimation test consistently used this strategy during the interview. For other students, this strategy tended to be a last resort when no other approach was recognized. Students had little success with this strategy partly because the interview items were generally very tedious and partly because of their own lack of skill to handle paper/pencil algorithms mentally. In several cases, it was difficult to tell whether the student was a good estimator or not, because the student was attempting to compute mentally rather than estimate on nearly every item.

*Conceptual understanding of percent, decimals and fractions.* The use of benchmarks (key reference points used as bounds in forming an estimate) on the two interview items dealing with percent, # 6 (30% of 54,215), and # 9 (15% of 28,000) highlights the conceptual understanding displayed by these Mexican students with regard to percent. One student utilized the benchmark strategy this way, "100% is 54,215, so 50% is about 25,000. 30% is less, about 18,000." When asked if he could do the problem another way, he responded, "You could find 1 % then multiply it by 30 but it would be harder to do." Following the completion of item # 6 all students were asked to tell what they knew about the number 30% and to compare it to 1/2 and 1/4. All but one of the eight students responded that it was "less than 1/2 or 50%" and "more than 1/4 or 25%". Although most students recognized and verbalized these relationships after completing the task, they did not use this information as they estimated. For example, one student produced an initial estimate of 1500 by using a front-end strategy (30 × 5 = 150 then annexed a 0) then adjusted to 2000 for a final estimate. This student never recognized the order-of-magnitude error even though, when asked, he said that 30%, " ... can be presented as a fraction, it's about 1/4. It's also less than 1/2 but more than 1/4." This failure to reflect on the estimate was common.

Whereas conceptual understanding aided the process of estimating with percents, lack of conceptual understanding severely limited students' ability to estimate with decimals and fractions. Tables VIII and IX report a summary of the results of interview items # 3 (12/13 + 7/8), and # 4 (486 × 0.24).

On the fraction item 12/13 + 7/8 (item # 3) one student noticed the nearness of each addend to 1 and estimated the sum to be 1 3/4. It took almost 50 seconds for this student to notice and verbalize this special relationship. In fact, during the estimation test, where time was restricted to 12 seconds, this same student resorted to an algorithmic approach of adding numerators and denominators to produce an estimate of 19/21. For this student, and perhaps others, time was needed to carefully examine the values before an estimate was produced. On the other hand, seven of the eight students used their interview time to mentally apply the standard paper/pencil algorithm taught in Mexican schools, or an erroneous version of it, to form their "estimate." For example, after over two minutes of mental work one student gave an estimate of 193/104 (exact answer is 187 /104 or 1 83/104). She described her strategy as applying the school taught written algorithm ({*ad*+ *be*}/*bd*) mentally. She said that since she did it mentally, she might be a bit off. This student never gave any indication that she saw either addend as near 1 or her answer as near 2.

Similarly, on item #4 (486 × 0.24) none of the eight interviewees produced a reasonable estimate on either the estimation test or during the interview. The most common technique employed for this problem was to mentally apply a paper/pencil algorithm. This strategy proved unsuccessful in all cases. One of the follow-up probes for this item was, "When you see the decimal number

0.24, what do you think of?" Four students verbalized that 0.24 is about 1/4, however the other four students expressed their attention to the decimal point. One student responded, "I think about that when I finish my answer I have to move my decimal point." Another said, "I have to remember to separate two numbers to place the decimal point." Finally, a third student stated, "When I see decimals I think about how hard they are."

TABLE VIII
Summary of interview data for item # 3.

| Strategy | Frequency | Estimate(s) produced | Time (sec.) |
|---|---|---|---|
| Item: 12/13 + 7/8 | | | |
| Uses idea of benchmark (each addend near 1) | 1 | 1 3/4* | 49 |
| Applies correct paper/pencil algorithm (with and without computational errors) | 4 | 193/104* | 120 |
| | | 103/204 | 38 |
| | | 100/104 | 29 |
| | | 160/104* | 129 |
| Applies incorrect paper/pencil algorithm (mentally) | 3 | 20/20 | 50 |
| | | 90/100 | 32 |
| | | 19/21 | 15 |

* Denotes estimates within the acceptable interval established by the researchers for this estimation item.

TABLE IX
Summary of interview data for item # 4

| Strategy | Frequency | Estimate(s) produced | Time (sec.) |
|---|---|---|---|
| Item: 486 × 0.24 | | | |
| Front-end (24 × 4 or 2 × 4) | 1 | 960 | 69 |
| Compatible rounding (1/4 of 486) | 1 | 450 | 39 |
| Applied paper/pencil algorithm, 486 (mentally) × .24 | 4 | 98.64 | 37 |
| | | 2400 | 58 |
| | | 2000 | 25 |
| | | 1400 | 45 |
| Could not understand strategy verbalized | 2 | 27 | 56 |
| | | 35 | 57 |

None of the estimates produced were within the acceptable range of estimates established by the researchers for this estimation item.

*Flexibility and possession of alternative strategies.* Following the completion of items # 1, # 4, # 6 and # 9, students were asked, "Can you think of another way to make an estimate?" The most common response was to restate the strategy already used or some extension of it (e.g. add front 2 digits for more accuracy rather than just the front or lead digit) or to respond that an estimate could be made using paper and pencil. For item # 9 (15% of 28,000) which lends itself to multiple strategies, five students said they could not think of another way to make an estimate. Each of the other three students stated one alternative strategy. Occasionally a student exhibited flexibility by stating a variety of alternative techniques. For example, for item # 1 one student, after applying an averaging strategy (5 × 90,000), stated that you could group pairs of addends then add these pairs to produce an estimate (compatible grouping), add the front digits and determine the place value (front-end), or apply the averaging strategy by multiplying 5 × 100,000 then compensating down (variation of original strategy used). The ability exhib-ited by this student to formulate alternative estimation strategies demon-strates her number sense and is a primary characteristic of good estimators as described by Reys et al. (1982). However, flexibility of this sort was rare in these interviews. The same student who described these alternative strategies often preferred to mentally apply paper/pencil techniques when forming estimates. She was a capable mental computer but did not seem willing to use more efficient techniques when the task called for an estimate.

*General estimation processes.* The three general processes, namely reformulation, translation and compensation, characterized in earlier research with a United States sample (Reys et al., 1982) were observed in these interviews. Reformulation was evident as students truncated numbers to apply the front-end strategy. The one student who changed 0.24 to 1/4 in item # 4 was also exhibiting this process. As was mentioned earlier, rounding, the most common application of reformulation in the United States, was rarely used in these interviews.

Translation was exhibited by three students in item # 1 as they applied the averaging strategy. One of these same three students again exhibited the process of translation in item # 6 as he described his strategy as changing 30% to 1/3 then dividing 54,000 by 3. Translation is clearly a more sophisticated technique than reformulation, and although it was apparent in these interviews among six of the students, it was less frequently observed than reformulation.

Compensation was utilized by every student at one time or another, and by at least one student on every interview item with the exception of item # 4. For example, one student used the benchmark strategy (combined with front-end) to initially recognize that the division of $21,319,908 by 26 (item # 7) must be less than $1 million (compared 26 to 21) then adjusted the estimate to "less than 900,000" by mentally multiplying 26 by 9 and comparing the result to 213 (9 × 26 = 234). Finally, she stated a final estimate of 800,000. Whereas she had a mathematically produced rationale for the initial adjustment (less than 900,000), the final adjustment reflects her intuitive guess at a reasonable estimate.

One student clearly demonstrated and verbalized the process of interme-diate compensation (adjusting numerical values prior to them being operated on to systematically correct for error) on the first interview problem. He explained his strategy this way, "Make the first [addend] 90 and the next 90 - these make up for each other. Then keep the third 90 and the last 2 each 90 - they make up for each other. Then take 90 × 5 or 450 so it's 450,000."

*Reflecting on estimates.* Consistent with parallel investigations with Japanese and United States students, these Mexican students rarely reflected on their estimates through their own initiative and rarely recog-nized unreasonable estimates. The occasions when they did change their estimate were when they were orally describing their strategies to the interviewer and realized an error. The rarity with which these students reflected on their estimates is consistent with similar findings regarding student reflection on paper/pencil or calculator-produced exact answers. These students, perhaps the best qualified to identify unreasonable answers because of their comparative skill at estimation, apparently do not use this skill to reflect on answers.

## DISCUSSION

Limitations
One of the purposes of this study was to replicate previous research done in the United States and Japan in a country that differed from them both culturally and educationally, in order to help validate and generalize the framework that describes the processes, strategies and characteristics of good estimators. Some of the strategies and all of the general processes described previously were apparent in the interviews with Mexican students who were identified as good estimators.

This research further documented the difficulties of assessing students' estimation abilities. Our observations of students and our examination of the testing results suggest that many factors may have influenced the low scores of the Mexican students. Among these factors are the following:

- Some of the students did not understand what "estimate" meant. Many students tried to use known paper/pencil algorithms mentally to get an exact answer. This was reflected in the high number of "significant" digits that many students used to report their answers.
- Mexican students are not used to quizzes or other types of tests where the time allowed per item is very short.
- Mexican students are not accustomed to tests where there is no opportu-nity to review a problem. In the screening test, each problem was projected on the screen only once.
- Many students kept working on a problem for more time than allowed, thus missing several of the subsequent items.
- Many students became frustrated at the fast pace of the test and their inability to produce a response in the time allowed. Frustration and discouragement were expressed in many ways: sometimes students would just stop trying to answer any of the problems.
- Students had never before had a test administered with a slide projector. For most of them, this was the first time they had seen a slide projector in school.
- Mexican students are not used to the presence of visitors in their classes. The screening test was conducted by a team of three or more people, none of whom were part of the regular school staff.

Framework

The screening test was useful in identifying the best estimators among the eighth-grade sample. However, given the overall low performance on the Computational Estimation Test of this sample, the term "good" estimator should be used advisedly. Nevertheless, interviews with these students revealed some of the same characteristics, strategies and cognitive processes found among good estimators in countries as diverse culturally and educa-tionally as the United States and Japan (Reys et al., 1991). This research provides further validation of the general framework previously hypothe-sized to describe the estimation process (Reys et al., 1982). Each strategy used by those interviewed could be described within the Reys' categorization of strategies. No new strategies or processes were observed. Additional research is needed to refine and further develop the framework to describe computational estimation processes of specific groups of students (age, talent, mathematical performance, etc.). It might also be useful to use the framework to predict computational abilities and performance of students according to their developmental characteristics (Sowder and Wheeler, 1989).

*Implications*

The fact that the same characteristics of good estimators that were found in other countries, were also found in Mexican students, is important. This means that some experience gained in other countries in the teaching of estimation can be used in Mexico, and may be transported to other cultures. The low performance of students in the screening test, specifically in 5th grade, reflects the curricular emphasis on paper/pencil algorithms and exact answers. Low performance of fifth and eighth graders also documents that most students do not develop estimation strategies on their own without instruction.

This research shows that at the present time estimation is neither taught nor learned in school, where exact computation is the primary goal. Our review of the grades 3-9 textbooks in Mexico revealed that exact paper/pencil computation dominated the curriculum, and no instruction in ap-proximate computation was presented. If change is to be brought about, it must address not only the mathematics curriculum but teacher education as well. In regard to the mathematics curriculum, an effort must be made to provide a better balance for the computational strand in elementary school. Less time must be spent on traditional written algorithms, and careful attention must be given to systematically exploring mental computation and estimation. Mexico's highly centralized educational system can make it easier to introduce changes in the mathematics curriculum which is pre-scribed for the whole country. However, it is not enough that mental computation and estimation be prescribed. Teachers must have educational materials suitable to the teaching of these topics. Present textbooks, both for elementary as well as for secondary school, should be revised to reflect a proper balance of computational alternatives. Teachers should become familiar with different estimation strategies and should learn how to integrate them naturally during instruction. It is not suggested that estima-tion be taught as a separate unit, but rather that it be integrated systemat-ically to provide regular estimation experiences prior to instruction on written algorithms. Teachers should help students decide which computa-tional alternatives (mental computation, written computation or estima-tion) is appropriate in certain situations. Teachers must also help students develop a respect for approximate answers and recognize that in computa-tional estimation there is not just one "right" answer.

Additional Needed Research

Further research along several lines is needed. Improved assessment proce-dures are needed to obtain more valid measures of estimation performance. A test with tight time constraints may not be the only way to assess performance on computational estimation, and alternative methods should be explored.

It is important to investigate estimation processes across a broader range of ages. Such research could provide valuable insight into the growth and development of estimation skills and processes over a period of time. This study reports only on computational estimation in the school setting. Ethnographic studies contrasting estimation performance, as well as strate-gies used by students in "real world" versus "school" settings are needed (Carraher et al., 1985). Such research seems particularly appropriate in the Mexican culture where a majority of students do not complete formal education.

Mental computation provides a related and highly fertile research area. Baseline data regarding mental computation performance are needed, along with the strategies and processes used by students at different stages. Research designed to explore and identify relationships between mental computation and estimation is also needed.

Finally, it is hoped that this report will stimulate more discussion and action among researchers. Both computational estimation and mental
computation are important goals in mathematics programs. Recent research in the United States, Japan and Mexico suggest that much needs to be done internationally to address the attention given to computation techniques such as estimation and mental computation. More cross-cultural studies are needed to provide an accurate and broad perspective, and it is hoped that this research will encourage other researchers to conduct studies which further this growing knowledge base.

ACKNOWLEDGEMENTS

The authors would like to gratefully acknowledge the support provided by the University of Missouri Research Council, the Fulbright Fellowship Program, and the Centro de Investigación en Matemáticas. We also acknowledge the support given by Martha Fabiola Carrillo of the Secretaría de Educación Cultura y Recreación to make possible the collecting of data.

The following students of the Facultad de Matematicas provided invaluable assistance in discussing research methodology, collecting and reviewing data: Ramon Cedillo, Livier Guzman, Jose Luis Laguna, Jovita Lerma, Cesar Mirabal, Francisco Mirabal, Ruben Perez Lara, and Fernando Ponce.

Also special acknowledgement to Alicia de la Macorra who helped in many ways throughout the research, and to Richard Shumway, The Ohio State University, and Rita Barger, Hickman Mills School District, Independence, Missouri for their suggestions to this manuscript.

REFERENCES

Carpenter, T. P., Corbitt, M. K., Kepner, H. S., Lindquist, M. M., and Reys, R. E.: 1980, 'Results and implications of the second NAEP mathematics assessment: elementary school', Arithmetic Teacher 27(8), 10-12, 44-47.

Carraher, T. N., Carraher, D. W., and Schliemann, A. D.: 1985, 'Mathematics in the street and in the schools', British Journal of Developmental Psychology 13, 21-29.

Cockroft, W. H.: 1982, Mathematics Counts, Her Majesty's Stationery Office, London.

Conference Board of the Mathematical Sciences: 1989, Everybody Counts: A Report to the Nation of the Future of Mathematics Education, National Academy Press, Washington, D.C.

Japan Ministry of Education: 1989, Curriculum of Mathematics for the Elementary School, Printing Bureau, Tokyo.

National Council of Supervisors of Mathematics: 1989, 'Essential mathematics for the twenty-first century: the position of the National Council for Supervisors of Mathematics', Mathematics Teacher 82, 388-391.

National Council of Teachers of Mathematics: 1989, *Curriculum and Evaluation Standards for School Mathematics*, Reston, VA.

Reys, R. E.: 1986, 'Evaluating computational estimation', in H. L. Schoen and M. J. Zweng (eds.), Estimation and Mental Computation, 1986 Yearbook, National Council of Teachers of Mathematics, Reston, VA.

Reys, R. E.: 1988, 'Testing computational estimation', Arithmetic Teacher 35(7), 28-30.

Reys, R. E., Bestgen, B. J., Rybolt, J. F., and Wyatt, J. W.: 1980, Identification and Characterization of Computational Estimation Processes Used by In-School Pupils and Out-of-School Adults, Report No. 79-0088, National Institute of Education, Washington, D.C.

Reys, R. E., Bestgen, B. J., Rybolt, J. F., and Wyatt, J. W.: 1982, 'Processes used by good computational estimators', Journal for Research in Mathematics Education 13, 183-201.

Reys, R. E., Reys, B. J., Nohda, N., Ishida, J., Yoshikawa, S., and Shimizu, K.: 1991, 'Computational estimation performance and strategies used by fifth and eighth grade Japanese students', Journal for Research in Mathematics Education 22, 39-58.

Rubenstein, R. N.: 1985, 'Computational estimation and related mathematical skills', Journal for Research in Mathematics Education 16, 106-119.

Solana, F., Cardiel Reyes, R., and Bolaños, R.: 1981, Historia de la educación pública en México, Fondo de Cultura Económica, México.

Sowder, J. T. and Wheeler, M. M.: 1989, 'The development of concepts and strategies used in computational estimation', Journal for Research in Mathematics Education 20, 130-146.

# 22 CALCULUS FOR MIDDLE SCHOOL TEACHERS USING COMPUTERS AND GRAPHING CALCULATORS[22]

**Course Development Activities**
A special experimental section of calculus for middle school mathematics teachers was offered at San Diego State University in the spring semester of 1990. The goal was to provide a "lean and lively" [2] introduction to the major ideas of differential and integral calculus, with minimal prerequisites in algebraic computation, accessible to students with a limited mathematical background, and with a strong conceptual emphasis. The use of graphing calculators and computer software was integrated into the course. Each student checked out a graphing calculator (Casio fx-7000g) for the semester.

The general outline of the course was to develop the notion of limit, relate it to the concepts of differentiation and integration, and then to link these two major ideas through the Fundamental Theorem of Calculus. The limit concept was presented through a unit on limits of sequences, based on the use of scientific calculators. This work on limits was followed by an introduction to linear functions, slope, and applications; these ideas formed the foundation for the introduction of the derivative. The derivative concept was introduced based on the idea of velocity [3]. Then, various applications of the notion of derivative were discussed. The materials were adapted so that the use of graphing calculators would be appropriate. The power of the graphing calculator enabled students to use trigonometric and exponential

---

[22] Flores, A., & McLeod, D. (1990). Calculus for middle school teachers using computers and graphing calculators. In *Proceedings Third Annual Conference on Technology in Collegiate Mathematics*. Columbus, OH: The Ohio State University.

functions, as well as polynomial and rational functions, in solving various kinds of problems dealing with rates of change, maximum and minimum points, and other traditional topics from the calculus curriculum.

The notion of integral was introduced as the area under a curve, and then the relationship between the integral and the anti-derivative was investigated [1]. This outline of the course content was influenced by a number of theoretical considerations based mainly on research in cognition science. For example, we build up abstract concepts out of more concrete examples, such as using velocity as the introduction to derivative. We used analogical reasoning, such as talking about periodic phenomena, and then connecting that idea to the periodicity of trigonometric functions. We emphasized to students that our goal was to improve their ability to think conceptually about the ideas of calculus, not just to work on computational proficiency. Connections among important ideas were developed to enable students to see calculus as an integrated collection of important ideas, rather than as a disconnected set of procedural skills.

**Evaluation Data**

The evaluation of the course used questionnaires, paper-and-pencil assessment of student performance, and interviews to gather cognitive and affective data that were related to student achievement. Cognitive assessment focused mainly on higher-order thinking skills in the domain of calculus. Data from the experimental class were compared with data from a traditional calculus class.

Students in the experimental section were recruited out of the ranks of prospective middle school teachers. They were invited to enroll in a special section of Math 121, Calculus for the Life Sciences. There were 10 regular students. A group of 11 students was selected at random out of another section of Math 121 (one dropped out).

At the beginning of the semester students filled out a background questionnaire and took a pretest. Data from this assessment indicated that the students had similar mathematics backgrounds, although their majors were different. As Table 1 indicates, seven of the ten students in the experimental section were elementary education majors, and eight of the students in the control class were majoring in the life sciences.

Data on the students' performance on the pretest is summarized in Table 2. Scores of both classes were quite low on the six-point pretest, which dealt with linear and quadratic functions and their graphs.

Table 1. Background data on participating students

| Characteristic | Experimental | Control |
|---|---|---|
| Major | | |
|    Elementary Teaching | 7 | 0 |
|    Life Science | 1 | 8 |
|    Other | 2 | 2 |
| Mathematics Background | | |
|    College Algebra | 5 | 4 |
|    Intermed. Algebra | 5 | 6 |

Table 2. Pretest and posttest mean scores

| Assessment | Experimental | Control |
|---|---|---|
| Pretest | 2.0 ($n = 9$) | 0.8 ($n = 10$) |
| Posttest | | |
|   Graphing | 1.7 | 1.1 |
|   Max/Min | 1.1 | 0.1 |
|   Distance | 1.0 | 0.3 |
|   Total | 3.8 ($n = 10$) | 1.5 ($n = 9$) |

The final exams for the two classes were different, since the classes had covered different material. For example, the standard Math 121 deals only with the derivative and its applications, and the integral is not introduced until the second semester of the course. However, there were seven items on the final that were common to both final exams. These items dealt with graphing a cubic function, solving a traditional max/min problem, and describing the flight of a helicopter when the distance function is given. The results (again in Table 2) suggest that the students in the experimental section were correct on about twice as many items as the control students. Of course, one must keep in mind that although both groups were allowed to use calculators on the final, only the experimental group was provided with the more powerful graphing calculators. These two paper-and-pencil assessments were supplemented by interviews with students from both the experimental and control sections.

Table 3. Interview questions

---

Has Math 121 influenced your beliefs about mathematics?
Has Math 121 influenced your attitudes toward mathematics?
Did you use calculators or computers in Math 121? If so, how did they influence your performance? Were there any difficulties in using them?
What is a function?
What is a limit?
What is a derivative?
What is an integral? (Only for students in the experimental class.)
What are the important relationships among these "big ideas" of calculus?

---

Interviews were conducted with eight students in the experimental group and with six of the students in the control group. All interviews were tape recorded; in addition, the interviewer kept notes on each session. The questions asked in the interview are listed in Table 3.

Students in both classes reported positive changes in beliefs and attitudes toward mathematics as a result of Math 121. Students in the control class were slightly more likely to think of mathematics in terms of following rules, and students in the experimental class were slightly more likely to report that they liked math more after this class.

Students in both classes made use of technology, but students in the control class were generally limited to finding logs or doing arithmetic on a standard scientific calculator of their own. All students in the experimental section reported extensive use of their calculators in graphing activities, and about half also used the computer software graphing package (Master Grapher) that was made available to them. Students reported no serious difficulty in using these technological aids. Students in the experimental group were uniformly positive about the use of graphing calculators, just as the control group was positive about using their scientific calculators.

All students had difficulty describing clearly the important concepts of calculus. Students in the experimental class were more likely to describe functions as a relationship or correspondence, and students in the control group were better able to give examples of applications of functions (primarily biological). About half of the experimental group could say something about the idea of limit (a number that you get closer and closer to, reducing the error of approximation), but their explanations were always vague and uncertain. Students in the control class were even less able to express ideas about limits clearly. In both classes, students would claim that they could find the limit, given a problem, but that it was hard to express the idea in words. Most students in both classes talked about the derivative in terms of its applications (e.g., finding maximum and minimum points) rather than in terms of its definition. A smaller fraction of the students could describe the derivative as a way of finding the slope or the rate of change of the function.

The concept of the integral was covered only in the experimental section. Most students discussed the integral in terms of the area under the graph of the function, the only interpretation that they had seen. Most of the experimental group was also able to relate the integral and the derivative in terms of the fundamental theorem of calculus, but almost no students in either class could describe any connection between the concept of limit and the concept of derivative.

**Implications for Curriculum Development**
Students in the experimental class developed a reasonable ability to apply the notions of calculus in solving problems, and they showed some conceptual understanding of the major concepts of the field. The use of the graphing calculator appeared to be very helpful in assisting these students (who were weak in algebra skills) in developing some proficiency in calculus concepts. It is relatively easy to help students develop the skills they need to use the graphing calculator to solve problems that would otherwise remain beyond their level of algebraic competence. The students had a great deal of difficulty, however, in expressing themselves clearly regarding calculus concepts. More experience in writing about these concepts or in explaining these concepts to their fellow students could help improve student performance in this area.

It was a bit surprising that no student brought up the notion of velocity to describe the concept of derivative. Generally, the research suggests that the initial example of a concept tends to be quite powerful and remains an important influence on the students' thinking for some time. In this case, however, it appeared that this initial example of the concept was overtaken by the more powerful and general notion of rate of change, or the slope of the function. This unexpected result could be due in part to the extensive use of the graphing calculator; the slope of the graph became a very salient part of the students' study of functions, something that they could see. Velocity, of course, is not easy to visualize.

In summary, the experimental course was reasonable successful in using technology to help students with inadequate algebra knowledge as they tried to understand all the major concepts of calculus in one semester. We need to find new ways to encourage students to concentrate on developing the conceptual knowledge that is most important for prospective teachers of school mathematics.

**References**
1. Courant, R., & Robbins, H. *What is mathematics?* New York: Oxford University Press, 1978.
2. Douglas, R. G. (Ed.). *Toward a lean and lively calculus*. Washington, DC: Mathematical Association of America, 1986.
3. Sawyer, W. W. *What is calculus about?* Washington, DC: Mathematical Association of America, 1975.
4. Steen, L. A. (Ed.). *Calculus for a new century: A pump, not a filter.* Washington, DC: Mathematical Association of America, 1988.

# 23 THEY'RE OFF![23]

**Activity 1.** Tossing dice: small group activity

Let's simulate a horse race in the following way: 12 horses take part, numbered from 1 to 12. Two dice are tossed and the sum called. The horse with that number advances one place. The first horse to advance 9 places is the winner.

Every student should choose one horse number. Taking turns, toss the dice and call the sum. Fill out the chart showing the progress of each horse.

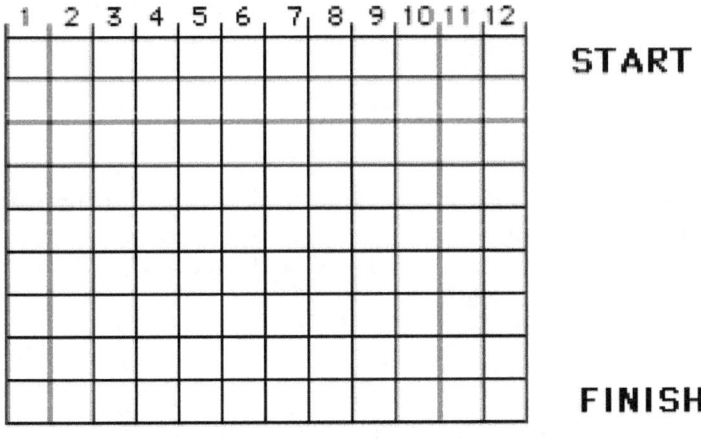

Chart 1

---

[23] Flores, A. (1990). They're off! In J. Trowell (Ed.) *Projects to Enrich School Mathematics Level 1*. National Council of Teachers of Mathematics, p. 72-78. Copyright National Council of Teachers of Mathematics. Used by permission.

Horse 1 is a lousy bet (why?), but any other horse at least has a chance to win.

**Activity 2.**

Compare your results with the results of other groups. Are horses in the middle (6, 7, 8) more likely to win than those on the sides (2, 3, or 11, 12)? Why?
Do you agree in your group that there is a favorite horse, that is, one that has a higher probability to win than the others?

Fill the chart below indicating in how many different ways a sum can be obtained. For example, the sum 4 can be obtained if the dice show 1, 3 or 2, 2 or 3, 1.

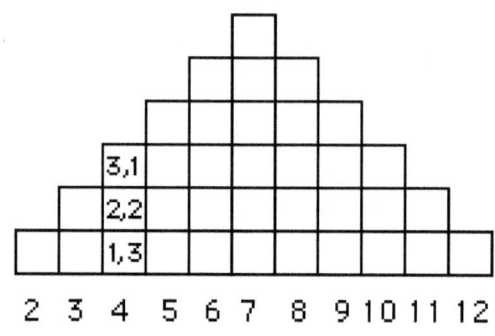

Chart 2

In how many ways can the sum 12 be obtained? _____ . In how many ways can the sum 7 be obtained? _____
According to the chart, what horse would be the favorite?
Does the favorite always win?

**Activity 3.** Use of random table

Let's simulate the race now using the attached random table.
Notice that in the table appear numbers between 1 and 6, grouped in pairs. The table simulates the results of tossing many pairs of dice.
To use the table, close your eyes and point with the pencil to some place of the table. Circle the pair of numbers closest to the mark. Afterwards, take the pairs of numbers as they appear on the table.
Sum the two numbers in each pair encountered. Proceed as in the previous

activity, filling out the chart showing the progress of each horse.

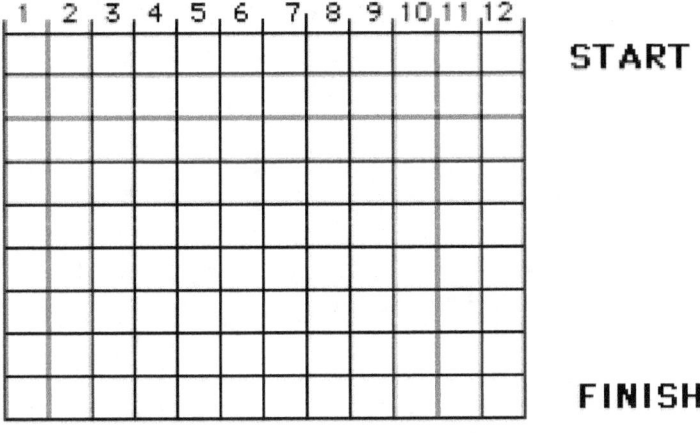

Chart 3

Choose a horse number and run the race. It takes less time using the random table than tossing the dice. Random tables are used to simulate random processes, without having to actually conduct the random experiment. Random tables can be constructed with the help of computers.

RANDOM TABLE
SIMULATION OF TOSSING 144 PAIRS OF DICE

```
2 4   3 6   1 4   2 1   2 3   5 2   6 4   1 2
5 2   3 6   1 5   5 4   3 1   5 5   3 4   1 2
1 4   1 1   6 5   2 4   3 1   3 6   4 1   1 5
5 5   2 1   6 6   4 5   4 6   6 3   4 4   3 1
1 2   4 1   4 1   2 6   3 5   5 4   6 5   5 3
5 4   4 3   3 4   3 5   5 6   6 2   6 6   1 4
1 4   1 1   1 2   6 4   2 4   1 6   3 3   4 2
5 6   6 5   6 6   5 3   4 6   3 2   1 1   1 5
3 2   4 1   5 5   6 2   4 2   3 3   3 2   2 6
3 3   6 3   2 2   3 5   5 6   6 2   4 1   4 1
4 2   2 6   3 2   5 2   2 4   1 6   4 2   1 4
1 6   4 1   1 1   1 2   4 2   6 3   5 1   2 2
5 6   5 6   6 1   5 4   4 2   5 1   4 4   4 5
1 2   3 1   4 3   6 1   5 6   3 3   3 3   5 2
2 2   1 1   2 5   3 2   6 6   6 4   4 1   2 2
5 1   3 3   5 1   4 4   5 4   5 6   1 6   3 6
4 4   5 5   5 2   5 3   1 6   4 6   5 2   3 1
5 3   6 5   2 2   2 2   2 4   1 3   4 5   3 4
```

**Activity 4.** Computer simulation
Here is a BASIC (Apple II version) program that simulates the race:

```
10 REM THEY'RE OFF!
20 HOME
30    DIM V(12)
40    LET R1 = 1 + INT (6 * RND (1))
50    LET R2 = 1 + INT (6 * RND (1))
60    LET S = R1 + R2
70    LET V(S) = V(S) + 1
80 VTAB (2*V(S)): HTAB (3*S): PRINT S
90 IF V(S) = 9 THEN END
100 GOTO 40
```

Choose a number again. RUN the program. Which horse was the winner? Run the program several times. Does the favorite horse always win? Is there an unexpected winner (a dark horse) sometimes?

What is the relationship between the shape of Chart 2 and the advancement of different horses?

Keep track of the winners in each race until a horse wins ten races.

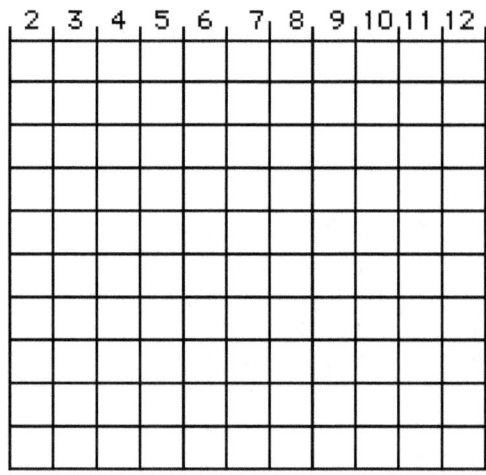

Chart 4   Winners in races of length 9

Write down how many times did each horse win:

# TO CONNECT IS TO UNDERSTAND MATHEMATICS 4

Horse:   2   3   4   5   6   7   8   9   10   11   12

**Activity 5.** Computer simulation

Modify line 90 to make the race shorter.

90 IF V(S) = 3 THEN END

Would you expect the favorite to win more often, or less often?

Run the program several times. Keep track of winners in each successive race in the following chart until a horse wins ten races.

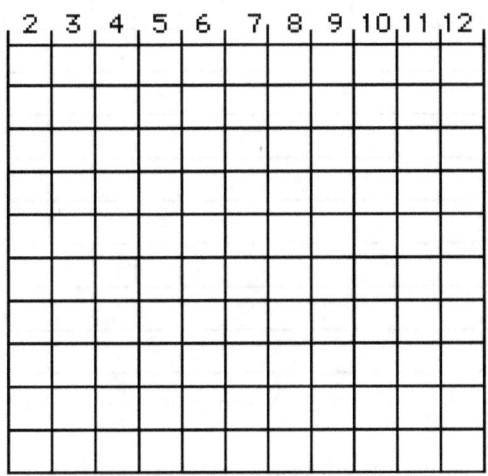

Chart 5   Winners in races of length 3

Write down how many times did each horse win:

Horse:   2   3   4   5   6   7   8   9   10   11   12

Compare the results of this activity with those of activity 4.

**Activity 6.** Computer simulation

Make the race longer. Change line 90.

    90 IF V(S) = 20 THEN END

What do you expect now? Will the favorite win more often, or less often?

You will have to change line 80 so that you do not run out of space on the screen.

    80 VTAB (V(S)): HTAB (3*S): PRINT S

Run the program several times. Keep track in the following chart of winners in each race until a horse wins ten races.

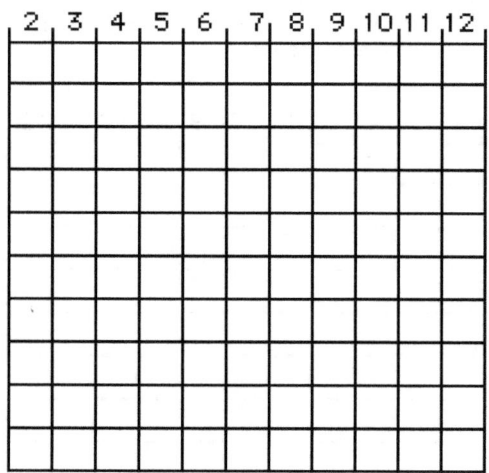

Chart 6. Winners in races of length 20

Write down how many times did each horse win:

Horse:   2   3   4   5   6   7   8   9   10   11   12

Compare the results of this activity with those of the previous activities. When can you predict with more certainty that a horse in the central lanes (6, 7, 8) will win, when the race is short or longer?

## References

Cwirko-Godycki, Jerzy. <u>Mathematical Activities from Poland</u>. Association of Teachers of Mathematics, (no place), 1979.

Flores, Alfinio. A microcomputer and the law of small mumbers. <u>Arithmetic</u>

<u>Teacher,</u> <u>31</u>(7), March 1984, 60-61.

Flores Peñafiel, A; Lerma Rico, J.; Martínez Cruz, A.; Mirabal G., F. <u>Prácticas de matemáticas para primero de secundaria</u>. Comunicaciones del CIMAT, 1987.

<u>Statistics and information organization</u>. Mathematics Resource Project. Preliminary Edition. Creative Publications, 1978.

# 24 EXPLORATION OF THE MEAN AS A BALANCE POINT[24]

**Grade levels:** 6 - 9

**Mathematics concepts, skills, processes**
Arithmetic average or mean, distance, midpoint

**Science concepts, skills, processes**
Lever, arm, weight, balance point, equilibrium

**Prerequisite skills**
Number line, directed distances, arithmetic average

**Objective**
Students shall develop the analogy of the mean of a set of numbers as the balance point of a lever with unit weights at those numbers on the number line.

**Rationale**
The arithmetic average, or mean of a set of data is a number that is used to summarize the information of all data. It is very helpful to deal with a single number that is representative, instead of dealing with many numbers. Providing different physical models of the mean can help students better understand and appreciate some of its properties.

---

[24] Flores, A., White, A. L. (1989). Exploration of the mean as a balance point. *School Science and Mathematics, 89*, 251-258. Used by permission of Wiley and School Science and Mathematics Association.

The lever, which has many applications in everyday life, can also be used to give physical meaning to the mean of a set of data. This type of physical thinking can be very fruitful in mathematics, for example, thinking in terms of levers helped Archimedes discover mathematical relationships such as the volume of a sphere and the area of a parabola.

The mean $\bar{x}$ of a set of data $x_1, x_2, x_3, ..., x_n$ is the sum of those numbers divided by the total number of data, that is,

$$\bar{x} = (x_1 + x_2 + x_3 + ... + x_n)/n$$

Students readily learn this computational formula for the mean. However, it is important for students to perceive the mean as a conceptual act as well as a computational act. Several concrete models can be used to develop the concept (Reys, Suydam & Lindquist, 1984). For example, children can use adding machine tape to represent scores, where the length of the strip is determined by the score. To show the average of two scores, tape the two strips of paper and then fold the resulting strip in half. Using columns of blocks and then asking students to "even-out" the blocks as much as possible is another concrete approach that can be taken to find the mean.

The following lesson outline highlights several ways of helping students grasp one physical interpretation of the mean. It begins with the simplest cases and then the model is extended to accommodate more data.

## Lesson Outline

### 1 Finding the mean of two numbers

If two children of equal weight want to balance on a see-saw, they have to sit symmetrically with respect to the position where the see-saw rotates. That is, they sit at the same distance from the point of balance, but in opposite directions (Fig. 1). The point of balance is the midpoint.

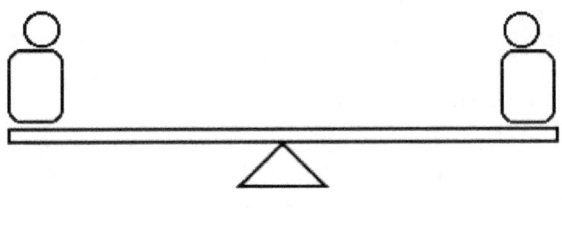

Fig. 1

The mean of two numbers *a* and *b*, is the midpoint of *a* and *b* on the number line (Fig. 2).

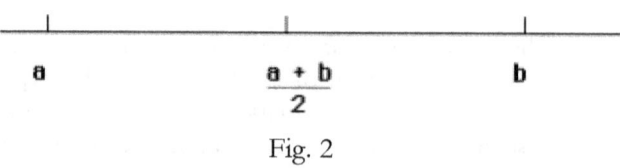

Fig. 2

For example, the mean of 5 and 9 is (5 + 9) / 2 = 7, and 7 is the midpoint of both numbers on the number line (Fig. 3)

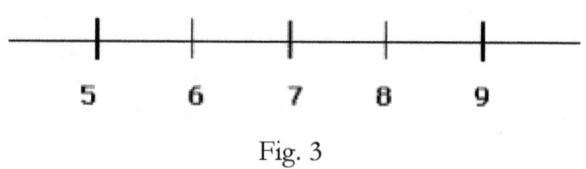

Fig. 3

The midpoint can be thought of as the point of balance for the two numbers in the following sense: imagine the number line as a weightless lever, and put a unit weight on the numbers we want to average. Then, if we hold the lever at the mean, the system is in equilibrium.
For example, the following system is in equilibrium (Fig. 4).

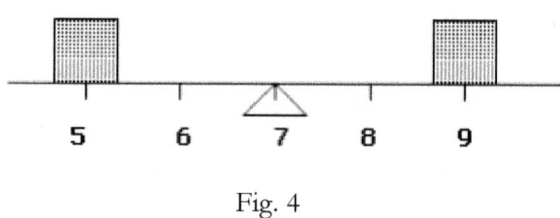

Fig. 4

The mean $\bar{x}$ of two numbers is at the same distance from both. Observe that $a - \bar{x}$ is the length of the arm of the lever form the point $\bar{x}$ to the point $a$ where the weight is located. Since both $a - \bar{x}$ and $b - \bar{x}$ represent distances that are equal in size but in opposite directions, they have different signs and their sum is zero. In the previous example, 9 - 7 = 2, 5 - 7 = -2, hence (5 - 7) + (9 - 7) = 2 + (-2) = 0. The mean $\bar{x}$ has the property that $(a - \bar{x}) + (b - \bar{x}) = 0$, since
$(a - \bar{x}) + (b - \bar{x}) = a + b - 2\bar{x} = a + b - 2(a + b)/2 = a + b - (a + b) = 0$

The sign of $a - \bar{x}$ tells us if the weight is to the left or to the right of the mean on the number line. If $a - \bar{x}$ is positive, the weight on $a$ tends to rotate the system counter clockwise, if $a - \bar{x}$ is negative, a weight on $a$ tends to rotate it clockwise (Fig. 5). Thus, the rotating effects of weights at $a$ and $b$ tend to cancel each other because $a - \bar{x}$ and $b - \bar{x}$ have different signs. The system is in equilibrium with respect to the mean because the sum of the rotating effects is zero.

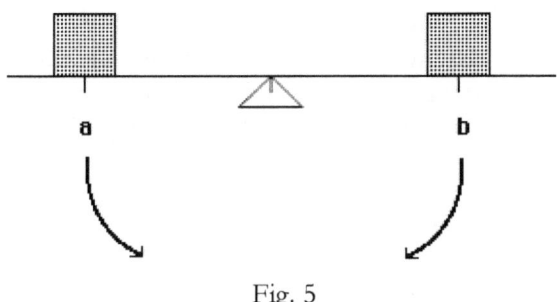

Fig. 5

The lever provides us with another model for additive inverses. If the mean of two numbers is zero, then the numbers are inverses of each other (Fig. 6).

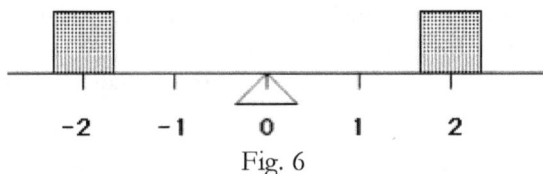

Fig. 6

For two equal numbers, the mean is the same number. In that case the two weights are located on the same point, and the point of balance is the same point.

## 2 The mean of three numbers

If we work with unit weights, the rotating effects with respect to $\bar{x}$ are given by the arm lengths $a - \bar{x}$, $b - \bar{x}$ and $c - \bar{x}$. If we have three equal weights, the system will be in equilibrium at $\bar{x}$, if the sum of the rotating effects is zero, that is, if
$(a - \bar{x}) + (b - \bar{x}) + (c - \bar{x}) = 0$ (Fig. 7)

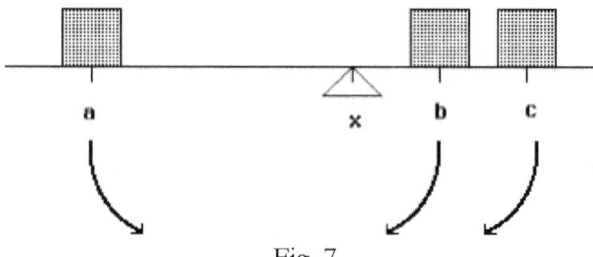

Fig. 7

If we think of the situation as a lever with three unit–weights located on the number line, balanced with respect to the equilibrium point, the mean $\bar{x}$ of three numbers $a$, $b$, and $c$ is precisely at the center equilibrium, since
$(a - \bar{x}) + (b - \bar{x}) + (c - \bar{x}) = 0$
This can be verified easily:
Let $\bar{x}$ be the mean of $a$, $b$, and $c$
$\bar{x} = (a + b + c)/3$
Then, since $a + b + c = 3\bar{x}$
$(a - \bar{x}) + (b - \bar{x}) + (c - \bar{x}) = a + b + c - 3\bar{x} = 0$
This means that the rotating effects of the weights at $a$, $b$, and $c$ cancel each other. Thus, if we hold the lever at $\bar{x}$, the system is in equilibrium.
For example, the numbers 2, 6 and 7 have a mean $\bar{x} = 5$. A weight at 2 tends to rotate the system counter clockwise $(2 - 5 < 0)$, the weights at 6 and 7 tend to rotate it clockwise $(6 - 5) > 0, 7 - 5 > 0)$.
Since $(2 - 5) + (6 - 5) + (7 - 5) = 0$, the system is in equilibrium (Fig. 8).

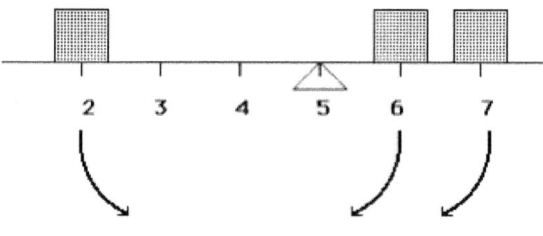

Fig. 8

Two of the numbers may be equal. In this case, two of the weights would be on the same point.

## 3 The mean in general
In general, if $\bar{x}$ is the mean of a set of data $x_1, x_2, x_3, ..., x_n$
we have that $x_1 + x_2 + x_3 + ... + x_n = n\bar{x}$, so that
$(x_1 - \bar{x}) + (x_2 - \bar{x}) + (x_3 - \bar{x}) + ... + (x_n - \bar{x}) = x_1 + x_2 + x_3 + ... + x_n - n\bar{x}$

$= 0$

Thus, if we think of the system as unit weights located at $x_1, x_2, x_3, ..., x_n$, and the number line as a lever balanced at $\bar{x}$, then the rotating effects $(x_1 - \bar{x}), (x_2 - \bar{x}), ..., (x_n - \bar{x})$ with respect to $\bar{x}$ cancel each other and the system is in equilibrium.

## 4 When is the mean not very representative?

The average of a set of data should be representative of the data. However, in some cases, when there is an extreme value, the mean might not be very representative: the extreme value can greatly affect the value of the mean. With the interpretation of the mean as the balance point of a lever, we can see why a single datum very different from the others has a big effect on the mean: that weight would have a very long lever arm.

For example, if the data are 1, 2, 3, 4, and 15, then the mean $\bar{x} = 5$ is not very representative since it is greater than any of the data except for 15 (Fig. 9).

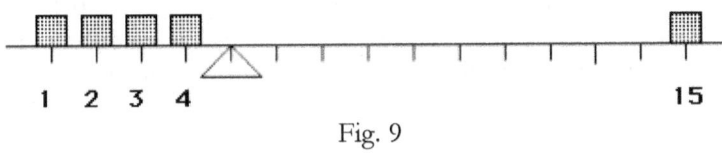

Fig. 9

Notice too, that by dropping the single datum point, 15, the mean changes dramatically. The new mean is 2.5 which is the midpoint of the remaining data. Because of this, other numbers that represent the data should be given in addition to the mean, for example the median or the mode, and also numbers that describe the dispersion of the data.

**Teacher notes:**

In the same way that a geometrical meaning can help to give students a better sense for certain numbers, for example areas and volumes for $n^2$ and $n^3$, the physical analogy of the mean as the center of equilibrium of the data can give students a better conceptual grasp of the mean. A balanced rod can be used to develop students' physical sense.

Researchers (Pollatsek, Lima and Well, 1981) have found that for many students, dealing with the mean is a computational rather than a conceptual act, and that knowledge of the computational rule does not imply any real understanding of the basic underlying concept. One possible way to develop the concept is by analog knowledge that might involve visual or dynamical images of the mean as a balance point. Strauss and Bichler (1988) list several properties of the mean that are important to understand it as a concept (in parenthesis are the analog properties of the balance point):

    a) the mean is located between the extreme values (the balance point is located between the extreme weights)

b) the sum of the deviations from the mean is zero (if the system is in equilibrium, the sum of rotating effects with respect to the balance point is zero)

c) the mean is changed by adding values other than the mean (if a weight is added not on the balance point, the equilibrium will be disturbed, and the new balance point will be different)

d) the mean does not necessarily equal one of the values that was summed (the position of the balance point is not necessarily at one of the weights)

e) the mean is representative of a group of individual values (the balance point is the center of equilibrium of all the weights)

The analogy of the mean as a balance point can contribute to a better grasping of these properties.

For readers interested in learning more about this topic, the books by Polya and Schiffer have nice discussions of levers. The analogy of the balance point can also be used for weighted averages. The analogy can be further extended for the mean of continuous distributions and centers of gravity.

**References**

FLORES PEÑAFIEL, Alfinio (1988). Giving physical sense to the average. *Comunicaciones del CIMAT.*

GARFIELD, J.; AHLGREN, A. (1988). Difficulties in learning basic concepts in probability and statistics: implications for research. *Journal for Research in Mathematics Educaction, 19*, 44-63.

HAACK, Dennis G. (1979). *Statistical literacy: a guide to interpretation.* Duxbury Press, 1979.

MATHEMATICS RESOURCE PROJECT (1978) *Statistics and information organization.* Creative Publications, 1978. "A balanced rod" p. 530.

MORONEY, M. J. (1956) *Facts from figures.* Penguin.

POLLATSEK, A.; LIMA, S.; WELL, A. D. (1981) Concept or computation: students understanding of the mean. *Educational Studies in Mathematics, 12*, 191-204.

POLYA, G. (1977) *Mathematical methods in science.* Mathematical Association of America.

REYS, R. E.; SUYDAM, M. N.; LINDQUIST, M.M. (1984) *Helping children learn mathematics.* Prentice Hall.

STRAUSS, S.; BICHLER, E. (1988) The development of children's concepts of the arithmetic average. *Journal for Research in Mathematics Education, 19*, 64-80.

SCHIFFER, M. M. (1984) *The role of mathematics in science.* Mathematical Association of America.

## 25 FAMILY PLANNING[25]

Mathematics Concepts/Skills
mean, random, ratio,
use of random table,
computer skills,
and prediction.

Science Concepts/Processes
population control,
data collection and
organization,
simulation, and inference.

Objective: Using simulation, the student will estimate the average size of a family in a community that uses a given method to plan the family.

Rationale: Simulating a phenomenon to predict its outcome is a process widely used when direct observation or experimentation is too expensive or not feasible. Coins, random tables, and computers can be used in school to simulate experiments.

Lesson Outline:

Materials: coin, random table or computer, student handouts.

Procedure:
Introduction
The community of Cihuatlan is patterned after a matriarchal system. It is therefore very important for a family to have a female descendant. The community has recently experienced a huge population growth due mainly to reduction in infant mortality and better life conditions in general. The community is concerned with the population explosion and has developed the following system to plan a family: each couple will have children until

---

[25] Flores Peñafiel, A. (1987). Family Planning. *School Science and Mathematics, 87* (6), central insert. Used by permission of Wiley and School Science and Mathematics Association.

they have a girl. After that, they will not have any more children. However, before implementing the system, they want to know what consequences it will have with respect to population growth, rate of males to females, average size of a family.

Student activity

1) Coin simulation
Simulate Cihuatlan's Family Planning System with a coin. Flip a coin until you get heads. Write T if you get tails, H if you get heads. Write the total number of times you have to flip the coin until you get heads.

Example:

```
Outcomes    Total

T T H       3
T H         2
T T T H     4
H           1
```

Write down the outcome of one experiment

Outcome Total
_____  ____

2) Use the random table to simulate the Family Planning System with ten couples.
This table has printed B or G in a random order. To use the table, close your eyes and pick a letter of the table. Circle that letter. From that place on, take the letters to the right in the order they appear on the table until you get a G. Begin another sequence of letters until you get a G. If you come to the end of a line continue with the line below. Repeat the experiment ten times.
Example: Suppose that on the line picked, letters appear in this order:

B G B B B   G G B B B   G B B B B   G B G G B   G B B G B   G B G G
B   B G G B B

The outcomes would be written in the following way:

```
B G         2
B B B G     4
G           1
```

| | |
|---|---|
| B B B G | 4 |
| B B B B G | 5 |
| B G | 2 |
| G | 1 |
| B G | 2 |
| B B G | 3 |
| B G | 2 |

Average: 2.6

Now write down the outcomes of your simulations and the total number of letters until you get a G. Take the average.

Outcomes    Total


Average:

Write down the averages of ten of your classmates:


Take the average of the averages: _____
This average represents the average of 100 simulations.
This average is an estimation for the number of children that a couple in Cihuatlan can expect to have if they follow the Planning System.

What is the expected value for the number of children per family? _____
How many boys and how many girls will there be? _____
What would the ratio of boys to girls be? _____
How fast would the population in Cihuatlan grow? _____

3) The random table was generated by a computer. You can also use a calculator or computer program to simulate the planning system.

```
PROGRAM:GIRL
ClrDraw
For(N,1,10)
1→U
While rand<.5
Text(6*N-6,6*U-6,"B")
1+U→U
End
Text(6*N-6,6*U-6,"G")
End
```

Run the program to see simulations for 10 couples. Run the program several times. The program can be easily modified to conduct more simulations, and take the average.

Conclusions.
A coin, a random table, or a programmable calculator can be used to simulate phenomena. In the case or Cihuatlan's Family Planning System we can take the results of many simulations to predict what would happen in the community if the plan was adopted.
If all couples in Cihuatlan follow the plan, the expected value for the number of children per couple is 2, and since there is always a girl, the ratio of boys to girls would be 1 to 1. According to this plan, Cihuatlan would have zero population growth.

Teacher Notes:
It is very instructional to flip the coins. Let all students simulate the Planning System with the coin at least once. Write down the results on the blackboard and discuss the outcomes.
The advantage of using the random table or the computer is of course that many more simulations can be done in less time and a more precise prediction made.
Extensions:
Students can also look at and graph the distribution of the outcomes. For example, about one half of the families would have only one child (Why?). About 1/4 would have two children, about 1/8 would have three, and so on.

## Reference
Flores Peñafiel, Alfinio; Lerma Rico, Jovita; Martínez Cruz, Armando; Mirabal García, Francisco. (1987). *Prácticas de matemáticas para segundo de secundaria*. Comunicaciones del CIMAT, Guanajuato, México.

# 26 PARABELLA[26]

Mother Quadratic was very happy with her new daughter, Parabella. The doctor was happy, too. He had feared a degenerate case, but fortunately Parabella was a round and healthy child. "She is beautiful," said Mother Quadratic. But what daughter is not beautiful in her mother's eyes? Parabella's elder sisters, Hyp and Elli, thought otherwise.

"Look how ugly she is. She has only one axis of symmetry," Hyp exclaimed (See fig. 1a).

"She has only one focus," Elli added. "She certainly is not like us!" (fig 1b, c).

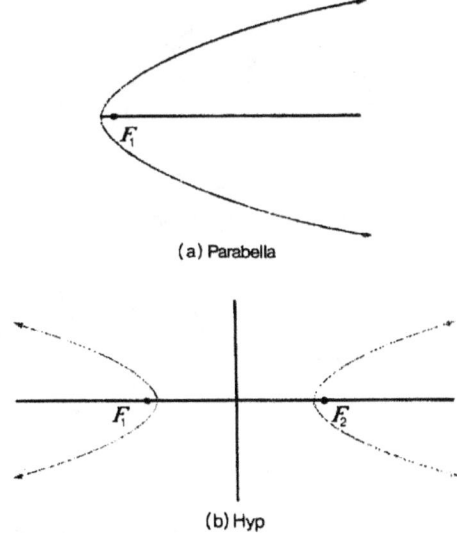

(a) Parabella

(b) Hyp

---

[26] Flores, A. (1985). Parabella. *Mathematics Teacher, 78*, 30 - 33. Copyright National Council of Teachers of Mathematics. Used by permission.

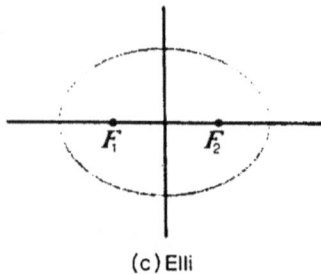

(c) Elli

**Fig. 1.** The three children of Mother Quadratic

When Elli and Hyp wore their Cartesian dresses, everybody could tell they were sisters:
Elli, $\frac{x^2}{a^2} + \frac{y^2}{b^2} = 1$ and Hyp, $\frac{x^2}{a^2} - \frac{y^2}{b^2} = 1$.
But Parabella looked different: $y^2 = 4px$.

Parabella was bright, and she soon discovered that her two sisters had other similarities. With Elli, the sum of the distances from each point to the foci was constant: $d_1 + d_2 = c$ (see fig. 2). For Hyp it was the difference of distances that was constant: $d_1 - d_2 = c$ (see fig. 3).

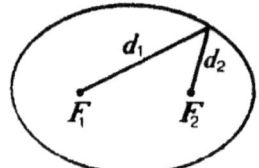

**Fig. 2.** $d_1 + d_2$ is constant.

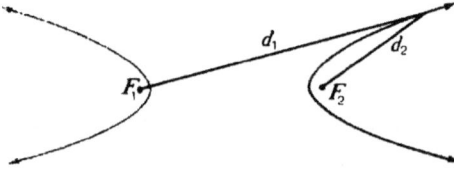

**Fig. 3.** $d_1 - d_2$ is constant.

Parabella was sad because she could not find a similar property in herself. She only had one focus. One day, when playing with a line, she made an amazing discovery. If she put the line perpendicular to her axis, so that the vertex was at the same distance from the line as it was from her focus, the same would be true for all her points (see fig. 5)!

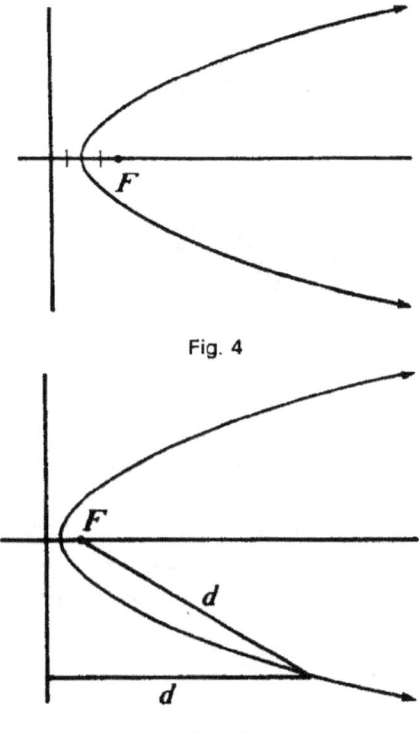

Fig. 4

Fig. 5

Parabella ran to tell her sisters the news.

"Look, with my friend directrix I have a property similar to yours!"

The two elder sisters responded coldly, "Who needs a stupid line? We don't need anything but our two foci."

Parabella was a little hurt. She wanted so much to convince her sisters that she was like them! The next day Parabella discovered that she could look at very distant objects. She was able to do this because she was able to concentrate parallel rays in her focus (see fig. 6). She had observed that Elli had a similar ability: Elli could concentrate the rays emitted from one focus in the other focus after reflecting them (see fig. 7).

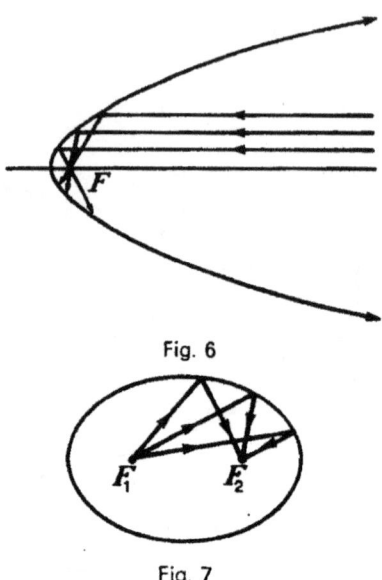

Fig. 6

Fig. 7

This time, however, Parabella did not tell her sister about her discovery. Parabella also discovered that this property of how the rays were reflected enabled her to remember how to get a tangent line, which she always forgot (see fig. 8). The same was true for Elli (see fig. 9).

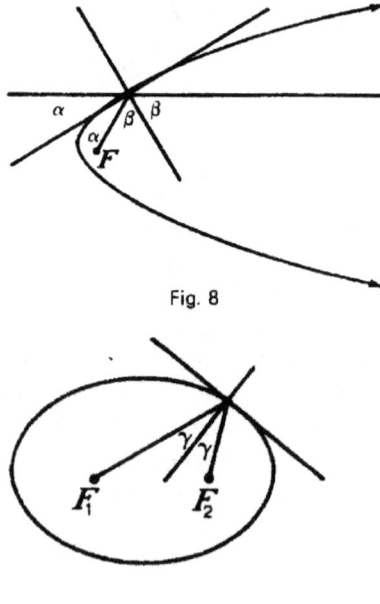

Fig. 8

Fig. 9

In the afternoon, Parabella read the book that her friend, the hexagon, had given her. It related the strange adventures of her friend's grandfather, a square. In the evening, Parabella ate lots of fraction pie and ice cream. She went to sleep and had a weird dream.

**Parabella's dream**
Parabella and her sisters were prisoners in Dandelin's castle. It was like a double ice cream cone (see fig. 10).

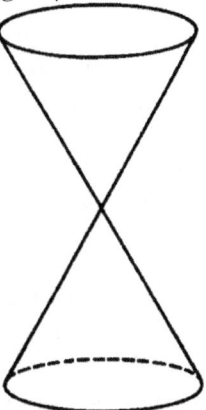

Fig. 10. Dandelin's castle

The castle was guarded by Dandelin's spheres. Parabella could hear her sisters, but she knew they were not in the same plane because the voices seemed to come from everywhere. One of the spheres was holding her at the focus. Two spheres held Elli tight (see fig. 11), and two more held Hyp. Prince Apollonius cut the cone with one of his planes, and Parabella managed to escape. Another slice freed Elli, and soon Hyp was free also (see fig. 12).

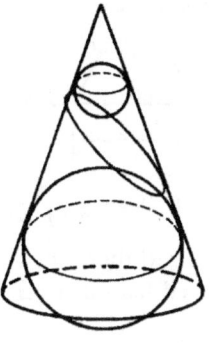

Fig. 11

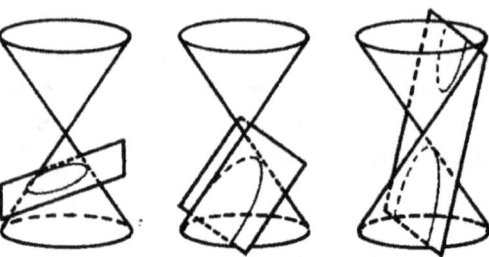

Fig. 12. The great escape

Parabella woke up very excited. She realized that she was more like her sisters than she thought. In one sense, she was between her sisters. The slice that set Parabella free was between the slice that freed Elli and the one that freed Hyp.

She ran to tell her sisters the dream. But as soon as she started to describe the castle, that cone in the third dimension, her sisters laughed at her. "That is what you get for eating so much at night and reading non-sense like *Flatland.*"

This time Parabella was very upset. "You and your planar thinking!" she cried and broke into tears. While she was weeping, the Polar Fairy appeared.

"Do not cry, Parabella. I will get. You some polar clothes and you will look like your sisters. You will also see that you are between them, as in your dream. I will even get a directrix for everyone."

The Polar Fairy did what she promised and soon the three sisters were dressed in their polar clothes (see fig. 13). The only difference among them was that —

for Elli: $e < 1$
for Parabella: $e = 1$
for Hyp: $e > 1$

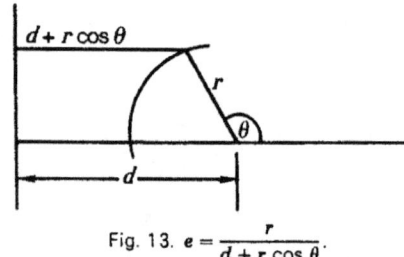

Fig. 13. $e = \dfrac{r}{d + r \cos \theta}$.

The elder sisters at last recognized that they all shared similar characteristics. Parabella was glad, and she lived happily ever after.

## Bibliography

Abbot, Edwin A. *Flatland: A Romance of Many Dimensions.* 2nd ed. Introduction by Banesh Hoffman. New York, Dover, 1984. Also 5th rev. ed., Harper

& Row, Publishers, 1963. (This is the classic book that caused Parabella's dream.)

Esbenshade, Donald H., Jr. "Adding Dimensions to Flatland: A Novel Approach to Geometry." *Mathematics Teacher* 73 (February 1983): 120-123.

Galbraith, Richard. "Mr. Normal Parabola." *Mathematics Teacher* 71 (January 1978): 36 - 38.

Johnson, Donovan A., and Gerald R. Rising. *Guidelines for Teaching Mathematics*, 2nd ed. Belmont, Calif: Wadsworth Publishing Co., 1972.

May, Kenneth O. *Elements of Modern Mathematics*. Reading, Mass.: Addiso-Wesley Publshing Co., 1959. (Gives a good treatment of conics where the unifying nature of eccentricity is emphasized.)

Ogilvy, C. Stanley. *Excursions in Geometry*. New York: Oxford University Press, 1969.

Toomer, G. J. "Apollonius of Perga." In *Dictionary of Scientific Biography*, edited by Charles C. Gillispie. New Yorl: Charles Scribner's Sons, 1970.

Whitt, Lee. "The Standup Conic Presents: The Parabola and Applications." *UMAP Journal* 3 (1982): 285-313.

# 27 A MICROCOMPUTER AND THE LAW OF SMALL NUMBERS[27]

If we throw a fair die, we know the probability of tossing any one of the six numbers is 1/6. In the long run, we expect each number to appear approximately in the same ratio. However, many persons do not have a sense of how long the "long run" is. Many people expect regularity even in the short run, what has been called the "law of small numbers" (Tversky & Kahneman, 1971). Writing short computer programs can help students to develop an understanding of random processes. Students themselves can write very short programs.

Program 1
10 DIM V(6)
20 LET N = INT(6*RND + 1)
30 LET V(N) = V(N) + 1
40 PLOT 2 * N, 2 * V(N)
50 IF V(N) = 10 THEN STOP
60 GOTO 20

Note: All programs run on the Timex-Sinclair 1000. For Apple computers, add
3 GR
4 COLOR = 13
to each program and use RND(1). For Radio Shack computers, change PLOT to SET(,).)

Program 1 simulates tosses of a die until one number appears ten times. Ask

---

[27] Flores, A. (1984). A microcomputer and the law of small numbers. *Arithmetic Teacher*, *31*(7), 60-61. Copyright National Council of Teachers of Mathematics. Used by permission.

## TO CONNECT IS TO UNDERSTAND MATHEMATICS 4

your students how many tosses will be made before the program stops. The answer should be between ten and fifty-five. (Why?) Let them plot the results of an imaginary experiment. Run the program a couple of times, and count the number of tosses. Compare the results with the guesses.

Since counting the tosses manually can be rather cumbersome, we can modify the program so that the computer counts the tosses.

Program 2
```
8 LET T = 0
10 DIM V(6)
20 LET N = INT(6*RND + 1)
30 LET V(N) = V(N) + 1
40 PLOT 2 * N, 2 * V(N)
45 LET T = T + 1
50 IF V(N) = 10 THEN STOP
60 GOTO 20
100 PRINT T
```

Table 1

Thirty-six tosses were needed in the first experiment before one number (a 1) was tossed ten times.

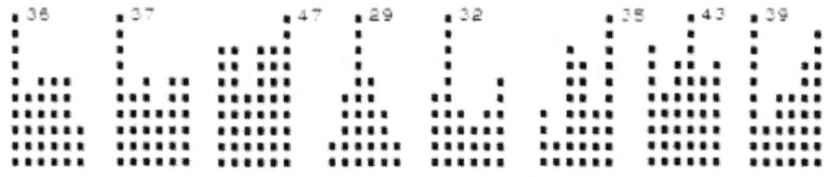

Some results of running program 2 several times appear in table 1. In the first run, thirty-six tosses were made before a 1 was tossed ten times. The fourth graph shows that twenty-nine tosses were made before a 3 was tossed ten times. Most students are quite surprised that the columns are so uneven. To explain this finding, ask the students to suppose that a teacher has thirty students. If the teacher asks ninety questions of individual students at random in one period, would each student be asked approximately the same number of questions? Program 1 can be easily modified to simulate this situation.

Program 3

```
10 DIM V(30)
15 FOR T = 1 TO 90
```

```
20 LET N = INT(30 *RND + 1)
30 LET V(N) = V(N) + 1
40 PLOT 2 * N, V(N)
50 NEXT T
```

The results of two runs of program 3 appear in table 2. Notice that one student received no questions (no mark in a column), whereas another was asked eight questions.

Table 2

---

Some students (in a class or thirty) are asked only a few questions, whereas one student in the first run is asked eight and another student in the second run is asked seven questions.

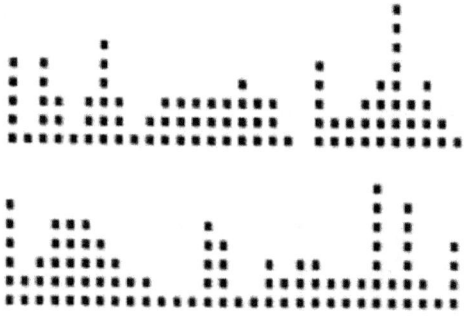

---

Modify program 3 to make sure every student gets one question. If the questions are asked strictly at random, you will be surprised how many questions it may take!

With a large number of experiments, we can develop a fairly good picture of a pattern of outcomes. If only a few cases are studied, however, we could be misled. Using a microcomputer is an excellent way to eliminate a belief in the law of small numbers.

Reference

Tversky, Amos, and Daniel Kahneman. "Belief in the Law of Small Numbers." *Psychological Bulletin* 76 (1971): 105-110.

## 28 OTAGATO[28]

Ever since he first arrived to Otagato, Alan felt a strange sensation. In the beginning, he did not know what it was. The same everyday actions, so familiar, left him with a feeling that they were different. They happened in the same way, and yet there was something that made him feel that all those things seen and done hundreds of times had something different. It was like that indefinable feeling of looking at one's own picture that is the same and at the same time different from the everyday image in the mirror.

At first he thought it was the change of place. Otagato's region was different. The intense blue sky made a stark contrast with the reddish soil, and the green of the prickly-pear and organ cactuses. Rocks would rise surprisingly in almost vertical ways. Otagato was a place where the light was more transparent and the wind always in movement gave a more tangible reality to everything. Far away objects seemed to be within the reach of the hand.
The first days in Otagato, Alan had felt disoriented. It was true that the place was prone to cause confusion. All streets seemed to have the name *Flood of 1905*. Because Otagato had been built on a ravine, it streets formed a complicated and indecipherable web for strangers. The street alleys bifurcated to infinity. It was impossible for any visitor to go from one place to another without passing twice by the starting point.

Otagato was a peculiar place and even more so its buildings. In the houses, it was common to find that a rock was the wall of the living room, or that houses were distributed in several levels, with terraces and stairs, garden in the upper level, and entrance through the roof. For example, in the guest house where he lived, to reach his room, he simply had to walk at the same level as the entrance of the house, but when he looked out his window he found out that he was at the second level.

---

[28] Flores Peñafiel, A. (1981). Otagato. In *Antología de relatos de ciencia ficción*. Guanajuato, Gto.: Universidad de Guanajuato.

The central building of the university was the most extraordinary. It seemed that the architect had used one of Escher's pictures of impossible perspective when doing the blueprint. At the university, after going through a maze of hallways and stairs, after climbing five or six floors he found himself suddenly at the street level.

But it was not those apparent paradoxes what made him feel that strangeness. Neither was the intricacy of the streets. What made him more uncomfortable was the fact that when walking through the streets and traveling along a circuit, houses would appear on the other side from where he expected to find them. For example, when going up along the street of Hemoglobin, the multi-family housing buildings were on the right side. If he then turned always to the left, he expected that the buildings would be on the outside of his trajectory. However, for some strange reason, the multi-family buildings appeared on the inside when he returned to them.

But he felt strange not only when walking through the streets. Some objects that he had seen many times seemed strangely alien: the snails in his garden, the small whirlpool formed in the water sink. Alan looked at them and did not quite figure what it was.
Later, the disturbing signs began to multiply. The street signs to indicate the routes for the tourists were inverted. Suddenly, when he saw that Red Cross van go by, and saw the sign ƎƆИA⅃UᗺMA he realized what it was. The objects that caused him the feelings of strangeness were like in a mirror. The whirlpool rotated in the opposite direction to the one he was used to. The snails had the spiral of their shells in the other direction. He also remembered that in the only street with a central divide in Otagato he almost had been run over by a car because the cars used the left side, like in England. However, in Otagato nobody seemed to pay attention to any of this.

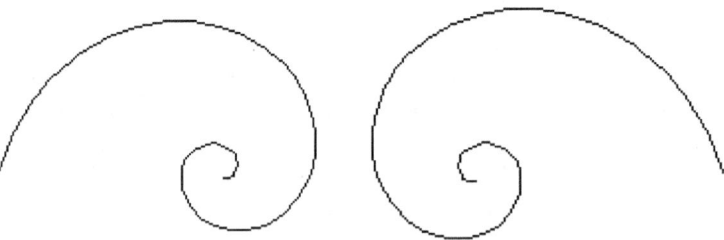

At school, his calculus teacher would write on the board using either his right or his left hand. Alan also realized, now in a conscious way, that the number of left-handed people in Otagato was surprisingly high. About half of the residents of Otagato wrote with their left hand, and the other half with their right. But what was most disturbing was that the same person would use one hand one day to write and the other the next day.

He also remembered that when he asked for directions in the street to find a place, the indications given by the residents of Otagato never included the words left or right. They simply used other kinds of references such as up, down, in front, straight ahead.

Alan tried to remember whether in any conversation someone from Otagato had ever used those words. He could not remember a single time. When he asked residents about left and right they did not understand, and he was unable to explain himself. The older inhabitants seemed to recognize the sound of the words. Words sounded familiar but they could not remember their meaning. Alan looked in newspapers and magazines published in Otagato and the words did never appear. They only appeared in the older books, those written before the year of the great drought, the year when the tunnel had been completed.

Alan thought that maybe the tunnel was the clue to the puzzle, but he could not find an explanation. He remembered vaguely a conversation he had with his friend Alfred about mathematics several years ago. At that time the conversation had to do with very strange mathematical objects. Curves that did not have tangent lines; closed bottles that did not have an interior, like Klein's; sequences of curves that became increasingly complex until the last one of them covered completely an area. Geometrical objects that were at odds with the world that was familiar to him. By some reason he associated the tunnel with that conversation.

Alan decided to write to Alfred, who was studying geometry at the University of Mayflower, doing research on surfaces and curves. He thought that maybe Alfred could explain the issue. He wrote detailing what was happening in Otagato. Alfred thought it was a joke. In any case, he decided to use it as an excuse to visit the city of Otagato, famous for its neo-colonial architecture and its mines.

Alfred arrived in the summer. Together with Alan he walked through the city. Although Alfred lost completely his sense of orientation and did not find an immediate explanation of why the street signs were inverted, he was still skeptical. He decided to go to the statue of Reyes, placed on one of the hills in the surroundings, to see from there the whole city. Alfred wanted to have a global image of the place. They arrived using the scenic highway, which because of its many curves, seemed the second-to-last step for a Peano curve. When looking at everything from above there was apparently nothing unusual, other than the extraordinary concentration of houses, built one on top of the next, and the web of streets and alleys that bifurcated and intersected themselves.

They decided to study the tunnel to see whether they could find the explanation. Alfred agreed that there they could find the clue to solve the mystery. When they investigated the history of how the tunnel had been built, they found that the excavation had been made by drilling simultaneously through opposite sides of the mountain, and the excavations were supposed to meet at the center. In that excavation, for the first time, the machines of centrifugal suction were used. The manufacturer indicated that the machines should not be used at their maximum power, because the centrifugal suction action distorted gravity and that could result in a turning effect. However, because the tunnel had to be inaugurated by the president on the planned day, the excavation was done on three shifts daily, working the machines at the maximum power. One of the machine operators told them how when operating the machine at its maximum power he had felt something very strange, something like a turn, some kind of rotation, but completely different to anything that he had experienced before. Because the operator could never explain what that sensation was, his friends would make fun of him, and so he ended up by not talking about the issue until he was interviewed by Alfred. The operator also told them that when the tunnel was inaugurated there was a big celebration in the city, in which absolutely everybody participated, from the youngest infants to the oldest people. With the tunnel would end the danger of the floods that every eleven years posed a threat to Otagato. Besides, the traffic of vehicles would be made much easier. The party lasted all night and the next day too. Part of the celebration was a walk with torches through the new tunnel circuit. Many became dizzy, but they thought it was because of the abundant wine they drank on that occasion. At the end, everyone fell asleep. When they woke up, the people felt a strange sensation, as if they were looking at things through a mirror. Gradually they all became used to it, and in a few days, they had all forgotten about it.

Alfred realized what had happened. His knowledge of non-orientable surfaces clearly led him to understand the situation. Alfred knew very well the Moebius strip. This two-dimensional surface has very strange characteristics, for example, that it only has one side. One can build a Moebius strip with a strip of paper. The strip is turned and point A is joined with point A', and point B with B'.

One can travel along the Moebius strip with a pencil and without lifting it from the paper travel throughout the whole surface. If we imagine two-dimensional beings that populated this strip, for them it would be impossible to see how the turn was made.

The mathematical explanation was obvious. During the tunnel construction, it had been rotated in a fourth dimension due to the effect of centrifugal suction. When the two excavations starting from the opposite sides of the mountain met at the center, Otagato became the three-dimensional analogous of a Moebius strip. But, how could he explain to his friend the logical conclusion? How was it possible that the centrifugal suction would distort space so much that the whole excavation had made a turn in the fourth dimension? Although the explanation was clear from a mathematical point of view, Alfred did not want to make it public because he could not explain in physical terms what was that fourth dimension in which the turn had taken place.

In any case, Alfred had spent a very nice vacation in Otagato and it was time to return to Mayflower. Alfred said goodbye to his friend and boarded the airplane. But next day, when he opened his geometry book he read the following.

## GEODESIC CURVES IN NON-ORIENTABLE SURFACES

## ABOUT THE AUTHOR

**Alfinio Flores** is professor of Mathematics Education at the University of Delaware. He has Mathematics degrees from the National University in Mexico and in Mathematics Education from Ohio State University. He teaches both mathematics methods and content courses *con ganas*. He uses technology, multiple approaches and concrete materials to develop conceptual understanding of mathematical ideas. He has published over 120 articles about the teaching and learning of Mathematics. He has conducted workshops for students ranging from Kindergarten to College, and taught courses for pre-service and in-service teachers in 32 states in two countries.

www.ingramcontent.com/pod-product-compliance
Lightning Source LLC
Chambersburg PA
CBHW050202230526
45470CB00001B/199